Summers, W. Kelly

Isotopes of water

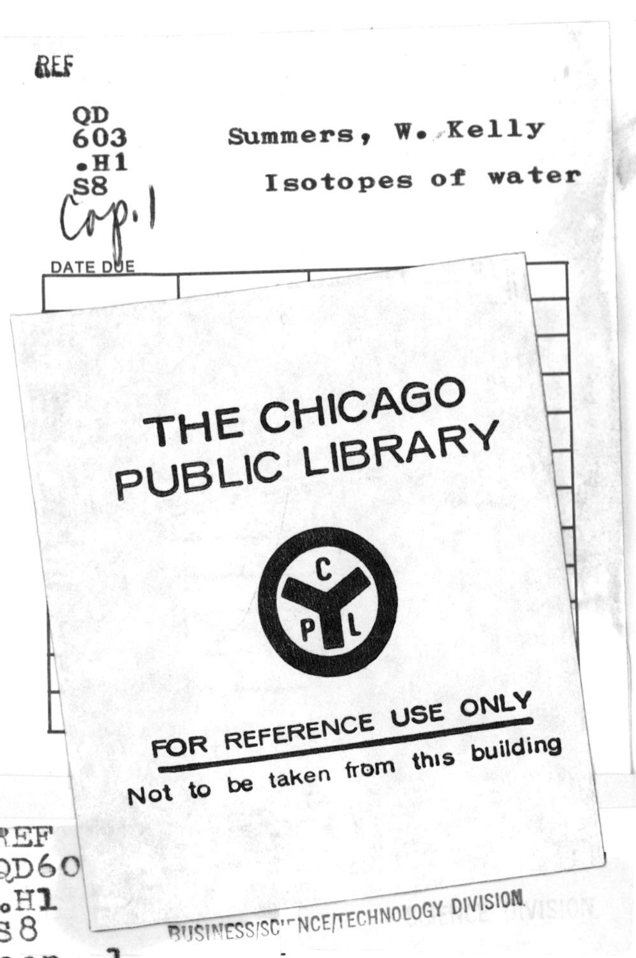

ISOTOPES OF WATER
A Bibliography

by

W. K. Summers

and

Carolyn J. Sittler

W. K. Summers and Associates
Consultants in Geological Fluid Dynamics
Socorro, New Mexico

P.O. Box 1425, 230 Collingwood, Ann Arbor, Michigan 48106

Library of Congress Catalog Card Number 76-1725
ISBN 0-250-40117-7

Manufactured in the United States of America

PREFACE

The isotopes of water as used here are hydrogen, deuterium, tritium, oxygen-16 and oxygen-18. Since 1950, the literature has increased from fewer than ten publications a year to more than 300 in 1974.

Tritium dominates the early literature. In recent years, the literature on the stable isotopes apparently has grown exponentially. A large part of the increased study of tritium can be attributed to detonation of the hydrogen bombs during the 1950s. These bombs released extraordinary quantities of tritium (radioactive hydrogen) to the atmosphere. Scientists became tritium-conscious, and the tritium concentration in water became a useful hydrologic tool.

The value of the stable isotopes became apparent when scientists realized that the ratios (D/H) and (O^{18}/O^{16}) in meteoric waters are temperature-dependent. These ratios are being used increasingly in the study of hydrologic systems and of rock-water interactions. As a consequence, we have included the references on hydrogen and oxygen isotope concentration in all natural sources—rocks, brines, inclusions, interstitial—as well as water that is obviously in the hydrologic cycle. We have also separated out those references relating to the prediction or interpretation of paleoclimatology and paleotemperature.

This bibliography started as a literature search to satisfy our need to know the significance of the isotopic concentrations reported for ground water and geothermal fluids. As we gathered reports, papers, books and documents, the range and diversity of the literature became clear. Hopefully, by presenting the organized list as we have here, we will shorten the time others will spend learning. Hopefully, too, this organized collection of references will help researchers recognize aspects of isotopic analyses and interpretations that need more work.

W. K. Summers
March 1976

ORGANIZATION

We have presented the references in two parts. Part I contains those relating to tritium; Part II contains those relating to deuterium and oxygen. In the interest of consistency, any reference that deals with both tritium and deuterium/oxygen is included in both parts. No other repetition occurs.

Within each Part references are included under the most appropriate category, although many references obviously belong in more than one. Many of these are relegated to a general category such as *Hydrologic Cycle*; others are arbitrarily assigned to a specific category. Therefore, we urge the user to scan not only the category of particular interest, but also related ones.

ORGANIZATION

We have presented the references in two parts. Part I contains those relating to tritium; Part II contains those relating to deuterium and oxygen. In the interests of completeness any reference that deals with both tritium and deuterium/oxygen is included in both parts. No other repetition occurs.

Within each Part references are indexed under the most appropriate category, although many references obviously belong in more than one. Many of these are referenced to a general category, such as Hydrology; others are arbitrarily assigned to a specific category. Therefore, we note the user should consult not only the category of particular interest, but also related ones.

SOURCES OF CITATIONS

1. Abstracts of North American Geology, 1966-1971
2. An Annotated-Indexed Bibliography of Geothermal Phenomena
3. Bibliography and Index of Geology, 1966-1975 (March)
4. Bibliography and Index of Geology Exclusive of North America, 1933-1968
5. Bibliography of North America Geology, 1950-1970
6. Geophysical Abstracts, 1951-1971
7. Geoscience Abstracts, 1959-1966
8. Meteorological and Geophysical Abstracts, 1950-1975 (March)
9. Selected Water Resources Abstracts, 1968-1975 (May)
10. Water Pollution Abstracts, 1957-1973
11. Water Resources Abstracts, Volumes 1-7

A cross check with a few volumes of other abstract journals such as *Biological Abstracts*, *Chemical Abstracts* and *Petroleum Abstracts* indicated that a comprehensive search of these sources would not yield enough new information to justify the effort.

ANNOTATIONS

The titles of most references cited describe their contents adequately. We have limited our supplementary remarks to a few terse words, and have annotated only to further clarify the content of the citation.

INTRODUCTION

This volume contains all the references on hydrogen (H), deuterium (H^2 or D), tritium (H^3 or T), oxygen-16 (O^{16}) and oxygen-18 (O^{18})—the isotopes of water—found during a concentrated library search of abstract journals and through corresponding with authors. It includes the references listed in a bibliography prepared for us by the National Technical Information Service. We believe it contains at least 95 percent of the references published or generally available in open-file reports through 1974, with a few references to 1975 publications.

In general, we have excluded the abstract references of those papers published later. References to abstracts, for the most part, mean that only the abstracts were published before 1975.

CONTENTS

PART I TRITIUM

PART II DEUTERIUM AND OXYGEN

PART I

TRITIUM

1. BIBLIOGRAPHY

Croxton, Fred Emory, 1950: Tritium (H^3) -- A bibliography of unclassified literature. U.S. Atomic Energy Comm. Rpt. TID-371, 32 p.

Feely, Herbert W. et al., 1961: The potential applications of radioisotopes technology to water resource investigations and utilizations. U.S. Atomic Energy Comm. NYO-9040, 340 p.

Contains an annotated bibliography

Foskett, A. C., 1961: Tritium detection and measurement. A bibliography. United Kingdom Atomic Energy Authority, Rpt. AERE-Bib 132, 36 p.

164 references

Hannahs, B. J. and Kershner, C. J., 1972: Bibliography on handling, controlling and monitoring of tritium (Dec. 1968-Jan. 1972). Mound Lab, Miamisburg, Ohio, Rpt. MLM-1946, 44 p.

Annotated

International Atomic Energy Agency, 1973: Isotope techniques in hydrology, v. 11 (1966-1971). IAEA Bibliogr. Ser. n. 41, 233 p.

814 references with abstracts

Rhodehamel, Edward C., Kron, Veronica B., and Dougherty, Verda M., 1971: Bibliography of tritium studies related to hydrology through 1966. U.S. Geol. Surv., Water Supply Paper 1900, 174 p.

Rudolph, A. W., Carroll, T. E., and Davidson, R. S., 1971: Tritium and its effects in the environment - A selected literature survey. Battelle Memorial Inst., Columbus Lab. Rpt. BMI-171-203, 378 p.

Skauen, D. M., 1963: Tritium in ecology - A review (in Proceedings 1st National Symposium on Radioecology, Colorado State Univ., Fort Collins, CO, Sept. 10-15, 1961). NY, Reinhold Pub. Corp. and Washington, D.C., Amer. Inst. Bio. Sci., p. 603-609.

U.S. Atomic Energy Comm., 1962: Special sources of information on isotopes. U.S. Atomic Energy Comm. Rpt. TID-4563, 3rd rev., 83 p.

648 references

Vaubert, B., and Bittel, R., 1970: Study of tritium hazards. Commissariat a l'Energie Atomique, CEA-BIB-182, EUR-4522F, 47 p.

Weed, H. C., 1972: Reactions of methane and carbon monoxide with tritiated water - a literature review. NTIS, UCRL-51238, 16 p.

2. SYMPOSIA, CONFERENCES AND COLLECTIONS

Anderson, L. J., 1965: Internationalt samarbejde. IAEA Dan. Geol. Foren., Medd., v. 15, pt. 4, p. 627-628.

Anonymous, 1970: International symposium on atmospheric trace constituents and atmospheric circulation (proceedings). J. Geophys. Res., v. 75, n. 15, p. 2875.

Barton, C. J., Butler, H. M., Cumming, R. B., and Rohwer, P. S., 1972: The 1971 tritium symposium at Las Vegas. Nuclear Safety J., v. 13, n. 3, p. 225-235.

Chatters, Roy M. and Olson, Edwin A., 1965: Proceedings of the sixth international conference on radiocarbon and tritium dating at Washington State Univ., Pullman, Washington, June 7-11, 1965. Pullman, WA, Washington State Univ., 800 p.

Columbia University Lamont Geological Observatory, 1957: Age measurements and other isotopic studies on Arctic materials. Final report Air Force contract 19(604)-1063, 6 p., entries paged separately.

Ferronskiy, V. I., Danilin, A. I., Dubinchuk, V. T., Polyakov, V. A., and Goncharov, V. S., 1968: Radioisotope investigation techniques in engineering geology and hydrogeology (Radioizotophyye metody issledovaniya, v inzhenernoy geologii i gidrogeologii). Izdatel'stvo Atomizdat, Moscow, 304 p; English translation, U.S. Atomic Energy Comm., Div. Tech. Info. Rpt. AEC-TR-7230 (TID-4500).

Hydrometeorological Services of the U.S.S.R., Moscow, 1974: Proceedings, all-union conference on nuclear meteorology, June 23-28, 1969, Obninsk. NTIS, TT-74-50011, 387 p.

International Atomic Energy Agency and the Joint Commission on Applied Radioactivity, 1962: Tritium in the physical and biological sciences. Proceedings of the symposium on the detection and use of tritium in the physical and biological sciences, May 3-10, 1961, Vienna (summaries in English, French, Russian and Spanish). Vienna, IAEA, v. 1, 369 p.

International Atomic Energy Agency, 1963: Radioisotopes in hydrology - an IAEA symposium. Nucleonics, v. 21, n. 5, p. 94.

International Atomic Energy Agency, 1963: Proceedings of symposium on the application of radioisotopes in hydrology, March 5-9, 1963, Tokyo (summaries in English, French, Russian and Spanish). Vienna, IAEA, 459 p.

International Atomic Energy Agency, 1967: Isotopes in hydrology - proceedings of the symposium on isotopes in hydrology, Nov. 14-18, 1966, Vienna, held by the IAEA in cooperation with the International Union of Geodesy and Geophysics. Vienna, IAEA, 740 p.

International Atomic Energy Agency, 1970: Isotope hydrology, 1970 - proceedings of symposium on isotopes in hydrology, 1970. Vienna, IAEA, STI-Pub. 255, 918 p.

International Atomic Energy Agency, 1971: Nuclear techniques in environmental pollution - proceedings of a symposium. Symposium on use of nuclear techniques in the measurement and control of environmental pollution, Oct. 26-30, 1970, Salzburg. Vienna, IAEA, 810 p.

International Atomic Energy Agency, 1974: Isotope techniques in ground water hydrology - proceedings of symposium, March 11-15, 1974, Vienna, Austria. Vienna, IAEA, 2 v.

International Geophysical Year, 1958: Annals, v. 5 (IGY instruction manual), pt. 4, 5 and 6, 1957/1958. London, Pergamon Press, p. 251-468.

Kruppner, Monica, ed., 1967: Isotopes in hydrology - proceedings symposium of International Atomic Energy and International Union Geodesy and Geophysics, Nov. 14-18, 1966, Vienna. Vienna, IAEA, STI-Pub. 141, 740 p.

Nelson, D. J., ed., 1971: Radionuclides in ecosystems, v. I - proceedings on the third national symposium on radioecology, May 10-12, 1971, Oak Ridge, Tenn. NTIS, CONF-710501, 678 p.

Olson, Edwin, A., and Chatters, R. M., 1965: Carbon-14 and tritium dating. Science, v. 150, n. 3702.

6th international meeting in Pullman, Washington, June 1965

Olsson, Ingrid U., 1966: C^{14}-Och H^{3}-mötet i Pullman, Washington, USA, 7-11 Juni, 1965. Geol. Fören. Stockholm Förh., v. 88, n. 2, p. 283-286.

Schell, W. R., and Sauzay, G., 1970: Use of isotopes in hydrology, Report on the 3rd international symposium, March 9-13, 1970, Vienna. Amsterdam J. Hydrol., v. 11, n. 4, p. 428-439.

Stout, Glenn, E., ed., 1967: Isotope techniques in the hydrologic cycle symposium, Univ. of Illinois, 1965. Am. Geophys. Union, Geophys. Mon. Ser., n. 11, 119 p.

Czachorski, Waclaw, ed., 1969: Applications of isotopes in geophysics, hydrogeology and engineering geology. Nukleonika, v. 13, n. 4-5, 224 p., English translation.

Renaud, A., et al., 1969: Etudes physiques et chimiques sur la glace de l'Indlandsis du Groenland 1959 (with English and German abs.). Exped. Glacial. Internat. Groenland, v. 5, n. 3 (Physical and chemical studies on the ice of the Greenland ice cap 1959). Medd. Grönland, v. 177, n. 2, 123 p.

3. METHODS FOR MEASUREMENT

Anonymous, 1973: Detection and Measurement (in Tritium). Phoenix, Arizona, Messenger Graphics, p. 86-191.

Anonymous, 1973: Environmental monitoring techniques (in Tritium). Phoenix, Arizona, Messenger Graphics, p. 496-556.

Anonymous, 1964: Report of the government chemist, 1963. London, Dept. Scientific and Industrial Research, H. M. Stationery Office, 124 p.

Describes techniques for measuring quantities of about 2 pc/ml

Anonymous, 1965: Report of the government chemist, 1964. London, Dept. Scientific and Industrial Research, H. M. Stationery Office, 116 p.

Enrichment methods

Allen, R. A., Smith, D. B., Otlet, R. L., and Rawson, D. S., 1966: Low level tritium measurements in water. Nucl. Instrum. Meth., v. 45, p. 61-71.

Anbar, M., Neta, P., and Heller, A., 1962: The radio-assay of tritium in water in liquid scintillation counters--the isotopic exchange of cyclohexene with water. Int. J. Applied Radiation Isotopes, v. 13, p. 310-312.

Armstrong, F. E., 1973: A low-level radioactive gas monitor for natural gas operations (in Proceedings of the Fourth Symposium on the Development of Petroleum Resources of Asia and the Far East, v. 2, pt. 2, documentation). U.N. Economic Comm. Asia Far East, Mineral Resources Div. Ser. n. 41, p. 146-149.

Portable detection system

Bainbridge, A. E., Sandoval, P., and Suess, H. E., 1961: Natural tritium measurements by ethane counting. Science, v. 134, n. 3478, p. 552-553.

Bainbridge, A. E., 1965: Determination of natural tritium. Rev. Sci. Instru., v. 36, n. 12, p. 1779-1782.

Barnes, J., Carswell, D. J., 1972: Measurement of low level environmental tritium samples. London J. Physics F. Sci. Instru., v. 5, n. 11, p. 1089-1090.

Barrata, E. J., 1973: Analysis for tritium in water--intercomparison study of November 1970; analytical control service, Winchester, Maryland (in Tritium). Phoenix, Arizona, Messenger Graphics, p. 522-530.

Birsoy, Y., 1972: Laboratory measurements of tritium in natural waters. Socorro, New Mexico, Institute of Mining and Technology, Masters Independent Study, p. 66.

Block, S., Hodgekins, D., and Barlow, O., 1971: Recent techniques in tritium monitoring by proportional counters. Univ. California, Lawrence Radiation Lab., Rpt. UCRL-51131 (NTIS), p. 21.

Broecker, W. S., 1959: Extension of technique for natural tritium measurements. Columbia Univ., Lamont Geol. Observatory, Final Rpt., Contract n. AF19(604)-4076, 25 p.

Bruner, F., Ciccioli, P., and DiCorcia, A., 1972: Gas chromatographic analysis of complex deuterated and tritiated mixtures with packed columns. Anal. Chem., v. 44, n. 6, p. 894-898.

Budnitz, R. J., 1974: Tritium instrumentation for environmental and occupational monitoring, a review. Health Physics, v. 26, n. 2, p. 165-178.

Burns, R. W., 1966: The enrichment and determination of tritium in water. Nucl. Sci. Abstr., v. 20, n. 233, 85 p.

Butler, F. E., 1961: Determination of tritium in water and urine. Anal. Chem., v. 33, p. 409-414.

Cameron, J. F., 1967: Survey of systems for concentration and low background counting of tritium in water (in Symposium on Radioactive Dating and Methods on Low-Level Counting, March 2-10, 1967, Monaco), Vienna, IAEA, p. 543-573.

Cameron, J. F., and Payne, B. R., 1966: Apparatus for concentration and measurement of low tritium activities (in International Conference Radiocarbon and Tritium Dating, 6th proc., 1965). U.S. Atomic Energy Comm. Rpt. CONF-650652, p. 457-470.

Cameron, J. F., 1966: A survey of systems for concentration and low background counting of tritium in water. Nucl. Sci. Abstr., v. 20, n. 692, p. 28.

Cantelow, H. P., et al., 1972: Sampling system for tritium oxide and carbon-14 in environmental air. Health Physics, v. 23, n. 3, p. 384-385.

Cooney, A., and Myatt, J. H., 1964: Liquid scintillation counting of Harwell effluent. United Kingdom Atomic Energy Authority, AERE-R 4678, 15 p.

Coplen, T. B., and Clayton, R. N., 1973: Hydrogen isotopic composition of NBS and IAEA stable isotope water reference samples. Geochim. Cosmochim. Acta, v. 37, n. 10, p. 2347-2349.

Curtis, M. L., 1972: Detection and measurement of tritium by bremsstrahlung counting. Int. J. Applied Radiation Isotopes, v. 23, n. 1, p. 17-23.

Dannecker, A. K., and Spittel, K. H., 1967: Measurement of tritium concentration in air (in Symposium on Instruments and Techniques for the Assessment of Airborne Radioactivity in Nuclear Operation, July 3-7, 1967, Vienna, Assessment of Airborne Radioactivity). Vienna, IAEA, p. 515-519.

Danilin, A. I., 1957: Primenenie iadernykh izluchenii v gidrometeorologu (Application of nuclear radiation in hydrometeorology), Leningrad, Gidrometeoizdat, 67 p.

Driver, G. E., 1956: Tritium survey instruments. Rev. Sci. Instru., v. 27, n. 5, p. 300-303.

Flanagan, F. J., 1970: Sources of geochemical standards pt. 2. Geochim. Cosmochim. Acta, v. 34, n. 1, p. 121-125.

Florkowski, T., Payne, B. R., and Sauzay, G., 1970: Interlaboratory comparison of analysis of tritium in natural waters. Int. J. Applied Radiation Isotopes, v. 21, n. 8, p. 453-458.

Hamilton, R. A., 1974: Determination of tritium in waste processing effluents by distillation and liquid scintillation emulsion counting. Atlantic Richfield Hanford Co. Rpt. ARH-SA-188 (NTIS), 24 p.

Hoffman, C. M., and Stewart, G. L., 1966: Quantitative determination of tritium in natural waters. U.S. Geol. Surv. Water Supply Paper 1696-D, 22 p.

Höhndorf, A., and Oro, F., 1971: Tritium fractionation in water-carbide to benzene reaction. Earth Planet Sci. Lett., v. 11, n. 4, p. 265-268.

Holmes, C. R. : Tritium. Socorro, New Mexico Inst. of Mining and Technology, Science Challenger, v. 1, n. 3, 6 p.

Hughes, B., Peiker, P., and Schumacher, E., 1969: Elektrolytische anreicherung des tritiums der Grönland-Eisproben (<u>in</u> Etudes physiques et chimiques sur la glace de l'indlandsis du Groenland 1959). Medd. Grönland, v. 177, n. 2, p. 42-54.

Hughes, M. L., and Holmes, J. W., 1971: Apparatus and method for the measurement of natural abundance of tritium. Adelaine, Australia Commonwealth Scientific and Industrial Research Organization, Div. Soils, Tech. Paper n. 3, 15. p.

Hunt, L. M., and Gilbert, B. N., 1972: Automated combustion versus digestion for tritium measurement in biological samples. Int. J. Applied Radiation Isotopes, v. 23, n. 5, p. 246-249.

Hutchins, W. H., Bradley, A. E., Morimoto, E. M., Wadleigh, W., and Willes, E., 1971: Facility for the determination of low-level tritium in biological and environmental samples. Univ. California Lawrence Radiation Lab Rpt. UCRL-73128 (NTIS), 10 p.

Kaufman, W. J., Nir, A., Parks, G., and Hours, R. M., 1962: Recent advances in low-level scintillation counting of tritium (in Tritium in the Physical and Biological Sciences), Vienna, IAEA, v. 1, pp. 249-261.

Khaytov, B. K., 1969: Izotopy v gidrogeologii (Isotopes in hydrogeology). Tashkent, Akad. Nauk Uzbek, SSR Inst. Yadernoy Geofiziki (Izdatel'stvo "FAN" Uzbek, SSR), 150 p.

Kinard, F. E., 1957: Liquid scintillator for the analysis of tritium in water. Rev. Sci. Instru., v. 28, p. 293-294.

Knowles, F. E., and Baretta, E. J., 1971: The determination of accuracy and precision limits for tritium in water. Radiological Health Data and Reports, v. 12, n. 8, p. 405-409.

Krishna Murthy, P. V., and Prahallada Rao, B. S., 1963: An all-metal spectrometer for the isotopic analysis of hydrogen at natural concentration levels. Indian J. Pure Appl. Physics, v. 1, n. 2, p. 73-79.

Lal, D., and Athavale, R. N., 1966: The measurement of tritium activity in natural waters. Nucl. Sci. Abstr., v. 20, n. 3819.

Lewis, J. D., 1972: A modified plastic bag combustion technique for the radioassay of C-14 and H-3 in biological tissues. Int. J. Applied Radiation Isotopes, v. 23, n. 1, p. 39-40.

Lindt-Muehlemann, C., and Oeschger, H., 1969: Messung der tritiumaktihität der angereich erten Grönland-Eisproben (in Etudes physiques et chimiques sur la glace de l'indlandsis du Groenland 1959). Medd. Grönland, v. 177, n. 2, p. 54-61.

McConnon, D., 1970: The use of water as a sampling medium for tritium oxide. Battelle Pacific Northwest Lab. Rpt. BNWL-CC547 (NTIS), 13 p.

Metson, P., 1969: A routine method for the determination of tritium by electrolytic concentration before liquid scintillation counting. Analyst, v. 94, n. 1125, p. 1122-1129.

Meyer-Wildhagen, F., 1972: Eine apparatur zur messung natürlicher kohlenstoff-14-und tritiumkonzen-trationen nach der gaszählrohmethode. Universität München.

Moden, D. D., Mullins, J. W., and Jaquish, R. E., 1973: Laboratory system for tritium analysis of large numbers of environmental samples (in Tritium). Phoenix, Arizona, Messenger Graphics, p. 512-516.

Moghissi, A. A., Bretthauer, E. W., and Compton, E. H., 1973: Separation of water from biological and environmental samples for tritium analysis. Anal. Chem., v. 45, n. 8, p. 1565-1566.

Moghissi, A. A., Bretthauer, E. W., and Carter, M. W., 1971: Monitoring of tritium under emergency conditions (in Proceedings International Symposium, July 5-9, 1971, Neuherberg, Rapid Methods for Measuring Radioactivity in the Environment), Vienna IAEA, SM-148-7, Conference 710705, p. 263-268.

Mullins, J. W., Andrews, V. E., and Dunn, L. M., 1972: Techniques for monitoring and analysis of environmental tritium. IEEE Transactions on Nuclear Science, v. 19, n. 1, p. 177-180.

Nishiwaki, Y., and Kauni, H., 1962: Measurement of the tritium concentration in natural waters. Shitsuryo bunseki, v. 10, n. 21, p. 83-90.

Ostlund, H. G., and Werner, E., 1962: Electrolytic enrichment of tritium and deuterium for natural tritium measurement (in Symposium on the Detection and Use of Tritium in the Physical and Biological Sciences), Vienna, IAEA, v. 1, p. 95-104.

Ostlund, H. G., Brown, R. M., Bainbridge, A. E., 1964: Standardization of natural tritium measurements. Tellus, v. 16, n. 1, p. 131-134.

Ostlund, H. G., and Mason, A. S., 1974: Atmospheric HT and HTO; Pt. 1, experimental procedures and tropospheric data 1968-72. Tellus, v. 26, n. 1 and 2, p. 91-102.

Ostlund, H. G., 1970: A rapid field sampling system for tritium in atmospheric hydrogen. NTIS TID-25345, 6 p.

Prydz, S., 1973: Summary of the state of the art in radiochromatography. Anal. Chem., v. 45, n. 14, p. 2317-2326.

Rieck, H. G., and Wogmar, N. A., 1968: Portable laboratory (in Annual Report 1967, v. 2, Physical Sciences, pt. 3, Atmospheric Sciences). Richland, Washington, Battelle Memorial Inst., Pacific Northwest Lab. Rpt. BNWL-715, p. 190-194.

Romanov, V. V., and Soyfer, V. N., 1961(1962): Apparatura i metodika izmereniya prirodnogo tritiya (Apparatus and procedure for measurement of natural tritium). Moscow, Gostoptekhizdat, p. 202-210.

Sauzay, G., and Schell, W. R., 1972: Analysis of low level tritium concentrations by electrolytic enrichment and liquid scintillation counting. J. Applied Radiation Isotopes, v. 21, n. 1, p. 25-33.

Schroder, L. J., 1971: Determination of tritium in water by the U.S. Geological Survey, Denver, Colorado. NTIS Geol. Surv. Rpt. USGS-474-134, 22 p.

Siksna, R., and Lindsay, R., 1961: Air ions produced by a tritium-ion generator, pt. 1, ion generators. Stockholm, Arkiv för Geophysik, v. 3, n. 6-12, p. 123-139.

Smith, D. B., and Rawson, D. S., 1961: Reconcentration of tritium by distillation (in Symposium on Detection and Use of Tritium in the Physical and Biological Sciences, May 3-10, 1961, Vienna). Vienna, IAEA, v. 1, p. 105-120.

Tamers, M. A., and Bibron, R., 1963: Benzene method measures tritium in rain without isotope enrichment. Nucleonics, v. 21, n. 6, p. 90-94.

Taylor, C. B., Polach, H. A., and Rafter, T. A., 1967: A review of the work of the Tritium Laboratory, Institute of Nuclear Sciences, New Zealand (in Nuclear Geology on Geothermal Areas, Spoleto, 1963). Cons. Naz. Ric. Lab. Geol. Nucl., p. 185-233.

Velten, R. J., 1972: Laboratory procedures (in Proceedings of Symposium on Direct Tracer Measurements of the Reaeration Capacity of Streams and Estuaries, July 7-8, 1970, Georgia Institute of Technology). EPA Water Pollution Control Research Ser. 16050 FOR 01/72, p. 55-65.

Wallick, E. I., and Knauer, G. A., 1969: Analysis of low level tritium by liquid scintillation counting without enrichment (abstr.). Amer. Geophys. Union Trans., v. 50, n. 4, p. 141.

Wallick, E. I., and Knauer, G. A., 1970: A modification of the benzene synthesis method for tritium analysis. Water Resources Res., v. 6, n. 3, p. 986-988.

Wolf, M., 1973: Benzolsynthese über Essigsäure zur low-level-messung von tritium. Int. J. Applied Radiation Isotopes, v. 24, p. 299-301.

4. DECAY RATE

Begemann, F., 1959: New determination of the natural terrestrial decay rate of tritium and the question of the origin of "natural" tritium (Neubestimmung der natürlichen irdischer Tritium zerfallsrate und die Frage der Herkunft des "natürlichen" Tritium). Zeitschr. Naturforschung, v. 14, n. 4, p. 334-342.

Jones, W. M., 1955: Half-life of tritium. Phys. Rev. v. 100, n. 1, p. 124-125.

12.262 ± 0.004 years

5. TRITIUM PRODUCTION

Anonymous, 1973: Tritium production (in Tritium). Phoenix, Arizona, Messenger Graphics, p. 30-85.

A review

Anonymous, 1958: Subcommittee on nuclear geophysics, cosmological and geological implications of isotope ratio variations. U.S. Natl. Acad. Sci., Natl. Research Council Pub. 572, Nuclear Sci. Ser. Rpt. 23, 187 p.

Fourth session includes a report on the distribution of radiocarbon and tritium and the production ratio of natural tritium

Aegerter, S. K., Loosli, H. H., and Oeschger, H., 1967: Variations in the production of cosmogenic radionuclides (in Radioactive Dating and Methods of Low-Level Counting, IAEA-ICSU Symposium, 1967, Monaco), Vienna, IAEA, p. 47.

Arnold, J. R., 1957: Production and distribution of radioactive isotopes in the atmosphere (in California Univ. Scripps Institute of Oceanography, Proceedings of the Conference on Recent Research in Climatology, March 1957), Univ. of California, p. 1-5.

Carach, J., and Cechvala, L., 1971: The sources of tritium and its radiation-hygienic problems. Prac. Lek., v. 23, n. 4, p. 120-125.

Craig, H., 1957: Radiocarbon and tritium distribution and mixing rates (in California Univ. Scripps Institute of Oceanography, Proceedings of the Conference on Recent Research in Climatology, March 1957), Univ. of California, p. 53-73.

Craig, H., 1957: Distribution, production rate and possible solar origin of natural tritium. Phys. Rev., v. 105, n. 3, p. 1125-1127.

Craig, H., and Lal, D., 1961: The production rate of natural tritium. Tellus, v. 13, n. 1, p. 85-105.

Fireman, E. L., 1953: Measurement of the (n, H^3) cross section in nitrogen and its relationship to the tritium production in the atmosphere. Phys. Rev., v. 91, n. 4, p. 922-926.

Grosse, A. V., Johnston, W. M., Wolfgang, R. L., and Libby, W. F., 1951: Tritium in nature. Science, v. 113, p. 1-2.

Korff, S. A., 1954: Effects of the cosmic radiation on terrestrial isotope distribution. Am. Geophys. Union, Trans., v. 35, n. 1, p. 103-106.

Lal, D., and Peters, B., 1967: Cosmic ray produced radioactivity on the earth (in Handbuch der Physik, Kosmische Strahlung, v. 2). Berlin, Springer-Verlag, v. 46, pt. 2, p. 551-612.

Lal, D., 1963: On the investigations of geophysical processes using cosmic ray produced radioactivity (in Earth Science and Meteorites). Amsterdam, North-Holland Publishing Co., p. 115-142.

Luianas, V. Iu., 1964: Otsenka skorosti obrazovaniia nekotorykh radioaktivnykh izotopov v atmosfere (Estimating the rate of formation of some radioactive isotopes in the atmosphere) (in Nauchnaia Konferentsiia po Iadernoi Meteorologii, February 3-6, 1964, Radioaktivnye izotopy v atmosfere i ikh ispol'zovanie v meteorologii). Moscow, Atomizat, p. 18-25.

Deals primarily with tritium

Martell, E. A., 1959: Artificial radioactivity from nuclear tests up to November 1958. U.S. Air Force Cambridge Research Center, GRD Research Notes, n. 19, 15 p.

Miskell, J. A., 1971: Production of tritium by nuclear weapons. Univ. of California Lawrence Radiation Lab. Rpt. UCRL-73270 (NTIS), 12 p.

Simpson, J. A., 1960: The production of tritons and C^{14} in the terrestrial atmosphere by solar protons. J. Geophys. Res., v. 65, n. 5, p. 1615-1616.

Teegarden, B. J., 1967: Cosmic-ray production of deuterium and tritium in the earth's atmosphere. J. Geophys. Res., v. 72, n. 19, p. 4863-4868.

Wilson, A. T., and Fergusson, G. J., 1960: Origin of terrestrial tritium. Geochim. Cosmochim. Acta, v. 18, n. 3-4, p. 273-277.

6. METEORITES AND MOON ROCKS

Bainbridge, A. E., 1962: The tritium content of three stony meteorites and one iron meteorite. Geochim. Cosmochim. Acta, v. 26, p. 471-474.

Begemann, F., 1962: Tritium determination in atmospheric gases and meteorites. Mainz, Germany, Max-Planck-Institut fur Chemie, Contract AF 61(052)-465, Summary Report n. 2, 7 p.

Begemann, F., 1966: Tritium content of two chondrites. Earth Planet Sci. Lett., v. 1, n. 4, p. 148-150.

Begemann, F., Eberhardt, P., and Hess, D. C., 1959: ^{3}He-^{3}H-strahlungsalter eines steinmeteorites (He^3-H^3 radiation age of a stone meteorite). Zeitschr. Naturforsch., v. 14a, n. 5-6, p. 500-503.

With English abstract

Bochsler, P., Eberhardt, P., Geiss, J., et al., 1971: Tritium in lunar material (in Lunar Science Conference, 2nd Proceedings, v. 2). Geochim. Cosmochim. Acta Suppl. n. 2, p. 1803-1812.

Buchwald, V. F., 1971: Tritium loss resulting from cosmic annealing, compared with microstructure and microhardness of six iron meteorites. Chem. Erde., v. 30, n. 1-4, p. 33-57.

Charalambus, St., and Goebel, K., 1962: Tritium and argon^{39} in the bruderheim meteorite. Geochim. Cosmochim. Acta v. 26, p. 659-663.

Charalambus, St., Goebel, K., and Stotzel-Riezler, W. Z., 1969: Tritium and argon^{39} in stone and iron meteorites. Naturforsch, v. 24a, n. 2, p. 234-244.

D'Amico, J., DeFelice, J., Fireman, E. L., et al., 1971: Tritium and argon radioactivities and their depth variations in Apollo 12 samples (in Lunar Sciences Conferences, 2nd Proceedings, v. 2). Geochim. Cosmochim. Acta, Suppl. n. 2, p. 1825-1839.

Fechtig, H., and Genther, W., 1965: Tritium diffusionsmessungen an vier steinmeteoriten (Tritium diffusion measurements on four stone meteorites). Zeitschr. Naturforsch., v. 20a, n. 12, p. 1686-1691.

With English abstract

Fesenkov, V. G., 1960: Osnounye dostizheniia meteoritiki za poslednee vremia (Principal achievements of meteorites in recent years). Moscow, Meteoritika, v. 18, p. 5-16.

Fireman, E. L., and Spannagel, G., 1972: Argon-37, argon-39, and tritium radioactivities in the Haveroe meteorite. Meteoritics, v. 7, n. 4, p. 554-564.

Fireman, E. L., D'Amico, J., DeFelice, J., et al., 1972: Radioactivities in Apollo 14 and 15 materials (in Lunar Science III). Lunar Sci. Inst., Contrib. n. 88, p. 262-264.

Fireman, E. L., and DeFelice, J., 1961: Tritium, argon-37, and argon-39 in the Bruderheim meteorite. J. Geophys. Res., v. 66, n. 10, p. 3547-3551.

Fireman, E. L., and DeFelice, J., 1960: Argon-39 and tritium in meteorites. Geochim. Cosmochim. Acta, v. 18, n. 3-4, p. 183-192.

Fireman, E. L., D'Amico, J., and DeFelice, J., 1970: Tritium and argon radioactivities in Apollo 12 samples and depth variations. Meteoritics, v. 5, n. 4, p. 197-198.

Fireman, E. L., D'Amico, J., and DeFelice, J., 1973: Depth variation of Ar^{37}, Ar^{39}, and H^3 in Apollo 16 material (in Lunar Science IV Abstracts), p. 248-250.

Fireman, E. L., and DeFelice, J., 1960: Argon 37, argon 39 and tritium in meteorites and the spacial constancy of cosmic ray. J. Geophys. Res., v. 65, n. 10, p. 3035-3041.

Fireman, E. L., DeFelice, J., and D'Amico, J., 1970: Depth variation of H^3, Ar^{37}, and Ar^{39} in lunar rock 12002. EOS, Am. Geophys. Union, Trans., v. 51, n. 7, p. 584.

Fireman, E. L., and Schwarzer, D., 1957: Measurement of Li^6, He^3, and H^3 in meteorites and its relation to cosmic radiation. Geochim. Cosmochim. Acta, v. 11, n. 4, p. 252-262.

Fireman, E. L., Fazio, G., and DeFelice, J., 1963: Argon 39, tritium and aluminum 26 in the Farmington meteorite and their discordant exposure ageslabs. Am. Geophys. Union, Trans., v. 44, n. 1, p. 89.

Fireman, E. L., 1962: Tritium in meteorites and in recovered satellite material (in Tritium in the Physical and Biological Sciences, v. 1, Symposium Proceedings, 1961). Vienna, IAEA, p. 69-74.

Fireman, E. L., 1961: Argon-37, argon-39, and tritium in recent meteorite falls (in Problems Related to Interplanetary Matter). Natl. Acad. Sci., Natl. Research Council Pub. 845 (Nuclear Sci. Ser. Rpt. 33), p. 28-30.

Fisher, D. E., 1967: Cosmogenic tritium problem in iron meteorites. J. Geophys. Res., v. 27, n. 4, p. 1351-1354.

Goebel, K., and Schmidlin, P., 1960: Tritium-messungen an steinmeteroiten (Tritium measurements on stone meteorites). Zeitschr. Naturforsch., v. 15a, n. 1, p. 79-82.

Kruger, S. T., and Heymann, D., 1968: Cosmic ray produced hydrogen 3 and helium 3 in stony meteorites. J. Geophys. Res., v. 73, n. 14, p. 4784-4787.

Lyman, W. J., Stoenner, R. W., and Davis, R., Jr., 1970: Argon and tritium radioactivities in lunar rocks. EOS, Am. Geophys. Union, Trans., v. 51, n. 7, p. 584.

Schultz, L., 1967: Tritium loss in iron meteorites. Earth Planet Sci. Lett., v. 2, n. 2, p. 87-89.

Singer, S. F., 1960: Production of tritium in nuclear spallations. Geochim. Cosmochim. Acta, v. 19, n. 3, p. 216-217.

Stoenner, R. W., Lindstrom, R. M., and Lyman, W., et al., 1972: Argon, radon and tritium radioactivities in the sample return container and the lunar surface (in Lunar Science, III, abstr.). Lunar Sci. Inst., Contrib. n. 8, p. 729-730.

Stoenner, R. W., Lindstrom, R. M., Lyman, W., et al., 1972: The cosmic ray production of ^{37}Ar, ^{39}Ar and tritium in lunar soil (in Lunar Science, III, abstr.). Lunar Sci. Inst., Contrib. n. 88, p. 731.

Stoenner, R. W., Lindstrom, R. M., Lyman, W., et al., 1972: Argon, radon and tritium radioactivities in the sample return container and the lunar surface (in Lunar Science, III). Geochim. Cosmochim. Acta, Suppl. v. 2, n. 2, p. 1703-1717.

Stoenner, R. W., Lyman, W., and Davis, R. Jr., 1971: Radioactive rare gases and tritium in lunar rocks and in the sample return container (in Lunar Sci Conference, 2nd Proceedings, v. 2). Geochim. Cosmochim. Acta, Suppl. n. 2, p. 1813-1823.

Stoenner, R. W., Lyman, W., and Davis, R. Jr., 1971: Argon and tritium radioactivities in lunar rocks and in the sample return container (in Lunar Sci. Conference). Houston, NASA, p. 223.

Tamers, M. A., 1963: Low concentration of tritium in iron meteorites. Nature, v. 197, n. 4864, p. 276-277.

Tilles, D., DeFelice, J., and Fireman, E. L., 1963: Measurements of tritium in satellite and rocket material, 1960-1961. Icarus, v. 2, n. 3, p. 258-279.

Trivedi, B. M. P., and Goel, P. S., 1969: Production of Na^{22} and H^3 on a thick silicate target and its application to meteorites. J. Geophys. Res., v. 74, n. 15, p. 3909-3917.

Wagener, K., 1967: Tritium loss from iron meteorites by solar wind hydrogen. Zeitschr. Naturforsch., v. 22a, n. 10, p. 1483-1488.

7. HISTORY AND STATE OF THE ART

Anonymous, 1973: Certain environmental aspects of tritium (in Tritium). Phoenix, Arizona, Messenger Graphics, p. 382-495.

Cumulative exposure index

Arnold, J. R., and Martell, E. A., 1969: The circulation of radioactive isotopes. Sci. Am., v. 201, n. 3, p. 84-93.

Role of nuclear devices

Bolin, B., 1960: On the use of tritium as a tracer for water in nature. Woods Hole Oceanographic Inst., Collected Repr. 1959, Contr. n. 1006, 8 p.

Deals with North America

Bolin, B., 1961: On the use of tritium as a tracer for water in nature (in International Geophysical Year 1957/ 1958). Annals, v. 11, p. 446-447.

Bowen, R., 1969: Progress in isotopic hydrology--the combined environmental isotope approach. Sci. Progr., v. 57, n. 228, p. 559-575.

Deuterium and oxygen-18 and carbon-14

Carlston, C. W., 1964: Use of tritium in hydrologic research--problems and limitations. Int. Assoc. Sci. Hydrology. Bull., v. 9, n. 3, p. 38-42.

Carlston, C. W., 1965: Reply to "Use of tritium in hydrology" by Bryan R. Payne (1965). Int. Assoc. Sci. Hydrology Bull., v. 10, n. 3, p. 67.

Carlston, C. W., and Thatcher, L. L., 1962: Tritium studies in the United States Geological Survey (In Symposium on the Detection and Use of Tritium in the Physical and Biological Sciences, 1961, Vienna). Vienna, IAEA, v. 1, p. 75-81.

Constantinescu, T., and Tenu, A., 1972: Application of environmental isotopes in Romanian hydrogeological research (incl. French, Russian summaries). Meteorol. Hydrol., n. 1, p. 37-55.

Czallany, S. C., 1966: Application of radioisotopes in water research (in Am. Water Resources Conf., 2nd annual, Chicago, 1966). Am. Water Resources Assoc., Proc. Ser. n. 2, p. 365-373.

Eriksson, E., 1962: Radioactivity in hydrology (in H. Israel and A. Krebs, eds., Nuclear Radiation in Geophysics). New York, Academic Press, p. 47-50.

Carbon-14

Eriksson, E., 1967: Isotopes in hydrometeorology (in Isotopes in Hydrology, Proceedings Symposium of International Atomic Energy Agency and International Union Geodesy and Geophysics, Nov. 14-18, 1966, Vienna). Vienna, IAEA STI/PUB/141, paper n. SM-83/2, p. 21-33.

Deuterium and oxygen-18

Gersht, E. P., 1957: Ob ispol'zovanii radioaktivnykh izotopov v meteorologii i gidrologii (On the use of radioisotopes in meteorology and hydrology). Leningrad, Meteorologiia i Gidrologii, n. 8, p. 62-64.

Grandelément, G., 1963: The use of radioactive isotopes as tracers in hydrology and hydraulics. Eau, v. 50, p. 293-300.

Jacobs, D. G., 1968: Sources of tritium and its behavior upon release to the environment. NTIS, TID-24635, 96 p.

AEC critical review series

Jouzel, J., and Merlivat, L., 1973: Le tritium dans le cycle natural de l'eau. BIST Commissariat à l'Energie Atomique, n. 178.

Koranda, J. J., Anspaugh, L. R., and Martin, J. R., 1972: The significance of tritium release to the environment. IEEE Trans. on Nuclear Science, v. 19, n. 1, p. 27-39.

Libby, W. F., 1959: Tritium in hydrology and meteorology (in P. H. Abelson, ed., Researches in Geochemistry). New York, John Wiley, p. 151-168.

Libby, W. F., 1966: Natural radiocarbon and tritium in retrospect and prospect (in Int. Conf. Radiocarbon and Tritium Dating, 6th Proceedings, 1965). U.S. Atomic Energy Comm., Rpt. CONF-650652, p. 745-751.

Magin, G. B., and Bizzell, O. M., 1965: Some applications of radioisotope technology to water resources investigations and utilization. Isotopes Radiat. Technol., v. 2, p. 124-133.

Mather, J. D., 1968: A literature survey of the use of radioisotopes in ground water studies. London, Natural Environment Research Council, Water Supply Papers of the Institute of Geological Sciences, Technical Communication n. 1.

Moghissi, A. A., and Carter, M. W., 1973: Environmental tritium (in Tritium). Phoenix, Arizona, Messenger Graphics, p. 399-405.

Mott, W. E., 1971: Isotopic techniques in the study and control of environmental pollution (in Symposium on Use of Nuclear Techniques in the Measurement and Control of Environmental Pollution, October 26-30, 1970, Salzburg). Vienna, IAEA, p. 3-46.

Nir, A., Kruger, S. T., Lingenfelter, R. E., and Flamm, E. J., 1966: Natural tritium. Rev. Geophysics, v. 4, n. 4, p. 441-456.

Tritium investigations, 1932-1966

Payne, B. R., Cameron, J. F., Peckham, A. E., Thatcher, L. L., 1965: The role of radioisotope techniques in hydrology (in Int. Conf. Peaceful Uses of Atomic Energy, 3rd proceedings, Geneva 1964), New York, United Nations v. 15, p. 226-238.

Special aspects of nuclear energy and isotope applications

Payne, B. R., 1965: Use of Tritium in hydrology (discussion of paper by C. W. Carlston, 1964). Int. Assoc. Sci. Hydrology Bull., v. 10, n. 1, p. 65-66.

Payne, B. R., 1967: Contribution of isotope techniques to the study of some hydrological problems (in Isotope Techniques in the Hydrologic Cycle). Am. Geophys. Union, Geophys. Mon. Ser., n. 11, p. 62-68.

Payne, B. R., 1972: Isotope hydrology. Advan. Hydrosci., v. 8, p. 95-138.

Comprehensive

Smierzchalska, K., 1973: Some aspects of investigation into pollution of the environment with radionuclides and heavy metals (in Polish). Postepy Naul Roln, v. 20, n. 5, p. 119-132.

Stewart, G. L., 1966: Experiences using tritium in scientific hydrology (in Int. Conf. Radiocarbon and Tritium Dating, 6th Proceedings, 1965). U. S. Atomic Energy Comm., Rpt. CONF-650652, p. 643-658.

Stewart, G. L., 1969: Tritium--a hydrologic tool. Annals of Arid Zone, v. 8, n. 2, p. 275-281.

Stewart, G. L., Wyerman, T. A., Farnsworth, R. K., 1968: Hydrological application of the tritium tracer technique. Am. Geophys. Union, Trans., v. 49, p. 166.

Thatcher, L. L., 1967: Water tracing in the hydrologic cycle (in Isotopic Techniques in the Hydrologic Cycle--Symposium 1965, Univ. Illinois). Am. Geophys. Union, Geophys. Mon. Ser., n. 11, p. 97-108.

8. PROCESSING AND HANDLING

Burger, L. L., 1974: Tritium separation and fixation. Richland, Washington, Battelle Pacific Northwest Lab. Rpt. BNWL-1825, p. 36-44.

Burger, L. L., and Ryan, J. L., 1974: The technology of tritium fixation and storage. Richland, Washington, Battelle Pacific Northwest Lab. Rpt. BNWL-1807 (NTIS), 36 p.

Dube, C. M., Coffin, D. O., and Stoll, R. D., 1973: An apparatus for the reduction of tritium emissions into the atmosphere. Los Alamos Sci. Lab., Rpt. LA-5303-MS (NTIS), 4 p.

Rhinehammer, T. B., and Lamberger, P. H., 1973: Tritium control technology. Mound Lab., Rpt. WASH-1269 (NTIS), 541 p.

Trevorrow, L. E., 1974: Storage of tritium and noble gases (in Argonne National Laboratory Waste Management Program Quarterly Report, October-December 1973). Argonne National Lab., Rpt. ANL-8089, p. 20-29.

Ward, W. E., 1973: Tritium control and technology. General Electric Co., Rpt. GEPP-104 (NTIS), 22 p.

Wethington, J. A., Jr., 1972: An investigation of tritium decontamination. Univ. California Lawrence Radiation Lab., Rpt. ORO-4201-2 (NTIS), 28 p.

9. PROPOSALS FOR USE

Brown, R. M., and Merritt, W. F., 1968: Tritium and other tracers in the hydrologic cycle (in International Conf. on Water for Peace, May 23-31, 1967, Washington, D.C.). U.S. Govt. Printing Office, v. 4, p. 618-626.

Craig, H., 1957: Isotopic tracer techniques for measurement of physical processes in the sea and the atmosphere. Washington, D.C., National Research Council Pub., n. 551, p. 103-120.

Carbon-14, deuterium, oxygen-13

Csallany, S. C., 1967: The application of isotope techniques in ground water hydrology (in Symposium on Ground Water Hydrology, 1967, San Francisco). Am. Water Resources Assoc. Proc. Ser. n. 4, p. 351-357.

Eriksson, E., 1963: Atmospheric tritium as a tool for the study of certain hydrologic aspects of river basins. Tellus, v. 15, p. 303-308.

Eriksson, E., 1967: Large-scale utilization of tritium in hydrologic studies (in Isotopic Techniques in the Hydrologic Cycle). Am. Geophys. Union, Geophysical Mon. Ser. n. 11, p. 153-156.

Harpaz, Y., Mandel, S., Gat, J. R., and Nir, A., 1963: The place of isotope methods in ground water research (in Radioisotopes in Hydrology). French, Russian and Spanish abstracts. Vienna, IAEA, p. 173-190.

Katsurayama, K., 1968: Tritium en hydrologie (concentration de tritium). Kyoto Univ. Research Reactor Institute, Annual Rpt., v. 1, p. 284-292.

Patin, S. A., 1965: Possibilities of using tritium for the study of dynamic processes in the ocean and in the atmosphere. Soviet Oceanography Acad. Sci. USSR Marine Hydrophys. Inst. Trans., 1963 ser., n. 2, p. 69-70.

Suess, H. E., 1969: Tritium geophysics as an international research project. Science, v. 163, p. 1405-10.

10. METHODS IN HYDROGEOLOGIC STUDIES

Clarke, W. B., and Kugler, G., 1973: Dissolved helium in ground water--a possible method for uranium and thorium prospecting. Econ. Geol., v. 68, n. 2, p. 243-251.

Tritium concentration used to date ground water

Davis, G. H., 1969: Use of hydrogen and oxygen isotopes in defining ground water flow systems (abstr.). Am. Geophys. Union Trans., EOS, v. 50, n. 11, p. 616.

Drost, W., Moser, H., Neumaier, F., and Rauert, W., 1974: Isotope methods in ground water hydrology. Brussels, Eurisotop Office Information Booklet n. 61, GSP-R56 (vgl. Nr.74), 176 p.

Eldor, M., and Dagan, G., 1972: Solutions of hydrodynamic dispension in porous media. Water Resources Res., v. 8, n. 5, p. 1316-1331.

Eriksson, E., 1958: The possible use of tritium for estimating ground water storage. Tellus, v. 10, p. 472-478.

Frohlich, K., Milde, G., and Hebert, D., et al., 1974: Methodische und messtechnische erkenntnisse ueber die anwendung von tritium, ^{14}C und ^{32}Si bestimmungen fuer hydrogeologische aufgaben (Techniques and measurements of tritium, C-14 and Si-32 applied in hydrogeology), incl. Russian and English summaries. Z. Angew. Geol., v. 20, n. 1, p. 16-21.

Geyh, M. A., 1969: Messungen der tritiumkonzentration in salzlauger. Kali und Steinsalz, v. 5, p. 208.

Geyh, M. A., and Mairhofer, J., 1970: Der natürliche carbon-14 und tritium gehalt der wässer. Steirische Beitrage zur Hydrogeologie, v. 22, p. 63-81.

Geyh, M. A., 1972: Die ^{14}C- und ^{3}H-methode in der angewandten hydrogeologie. Berlin, Wassenkalender, Erich Schmidt Verlag, p. 18-36.

Geyh, M. A., 1972: Basic studies in hydrology and ^{14}C and ^{3}H measurements (abstr.). Int. Geol. Congr. Abstr. - Congr. Geol. Int., v. 24, p. 340.

Geyh, M. A., 1973: Die anwendung der ^{14}C- und ^{3}H-methoden bei hydrologischen untersuchungen in der umselt forschung (The application of C-14 and H-3 methods for hydrological environmental studies). Dtsch. Geol. Ges. Z., v. 124, pt. 2, p. 515-522.

Günay, G., 1969: Yerati suyu arastirmalarinda izotoplardon ve diger izleyicilerden yararlanma (Ground water isotope studies). Türk. Jeol. Kurumu, Bül, v. 12, n. 1-2, p. 58-65.

Nir, A., 1967: Development of isotope methods applied to ground water hydrology (in Isotope Techniques in the Hydrologic Cycle, Univ. Illinois Symposium, 1965). Am. Geophys. Union, Geophys. Mon. Ser., n. 11, p. 109-116.

Padzik, A., Niemczynowicz, J., and Fraczek, E., 1969: Use of tritium in hydrogeological studies. Nukleonika (English trans.), v. 13, n. 4-5, p. 78-87.

Rauert, W., 1971: Ueber messungen von tritium und tritium und kohlenstoff-14 bei hydrologischen untersuchungen (The measurement of tritium and carbon-14 for hydrological studies). Geol. Bavarica, n. 64, p. 36-74.

Roth, E., Merlivat, L., Courtois, G., Cornuet, R., Guizeriv, J., Margrita, R., Molinar, J., Gras, R., 1971: Application des isotopes stables et radioactifs dans la domaine de l'hydrologie et de la sedimentologie. Geneva, 4th Int. Conf. of the U.N. on the Peaceful Uses of Atomic Energy, Sept. 6-16, 1971.

Simpson, E. S., 1973: Finite state mixing cell models. Am. Geophys. Union Trans., EOS, v. 54, n. 4, p. 259-260.

Von Buttlar, H., 1958: Investigating ground water by analysis of atmospheric tritium. J. AWWA, p. 1533-1538.

11. HYDROLOGIC CYCLE

Begemann, F., 1956: Distribution of artificially produced tritium in nature (in Nuclear Processes in Geologic Settings). Natl. Acad. Sci., Natl. Research Council Pub. 400, p. 166-171.

Begemann, F., and Libby, W. F., 1957: Continental water balance ground water inventory and storage times, surface ocean mixing rates and world-wide water circulation patterns from cosmic ray and bomb tritium. Geochim. Cosmochim. Acta, v. 12, n. 4, p. 277-296.

Begemann, F., 1958: The use of tritium in studies of hydrological problems (in International Geophysical Year, 1957/1958 Annals). IGY Instruction Manual, Pt. 5, p. 344-348.

Eriksson, E., 1961: Tritium and the circulation of water in nature. WMO Bull., v. 10, n. 4, p. 212-216.

Eriksson, E., and Bolin, B., 1965: Oxygen-18, deuterium, and tritium in natural waters and their relations to the global circulation of water (in Radioactive Fallout from Nuclear Weapons Test, USAEC Proceedings, November 3-6, 1964, Germantown, Maryland). Abstract.

Eriksson, E., 1967: The atmospheric transport of tritium (in Isotope Techniques in the Hydrologic Cycle). Am. Geophys. Union, Geophysical Mon. Ser., n. 11, p. 56-57.

Farnsworth, R. K., Sherman, M., and Stewart, G. L. Tritiated water budget for the United States 1963-1966, abstr. Am. Geophys. Union Trans., v. 49, p. 166.

Giletti, B. J., Bazan, F., and Kulp, J. L., 1958: The geochemistry of tritium. Am. Geophys. Union, Trans. v. 39, n. 5, p. 807-818.

International Atomic Energy Agency, 1967. Tritium and other environmental isotopes in the hydrological cycle. Vienna, IAEA, Tech. Rpt. Ser. n. 73, 92 p.

Kaufman, S., and Libby, W. F., 1954: The natural distribution of tritium. Phys. Rev., v. 93, n. 6, p. 1337-1344.

Taylor, C. B., 1966: Tritium in southern hemisphere precipitation 1953-1964. Tellus, v. 18, n. 1, p. 105-131.

Taylor, C. B., 1968: Comparison of tritium and strontium-90 fallout in the southern hemisphere. Tellus, v. 20, n. 4, p. 559-576.

Thatcher, L. L., 1967: Tritium fallout maps, implications to hydrologists (abstr.). Am. Geophys. Union, Trans. v. 48, n. 1, p. 99-100.

A. REGIONAL MONITORING OF NATURAL WATER

Anderson, K., 1969: Tritium in New York State waters. Radiological Health Data and Reports, Washington, D.C. v. 10, n. 3, p. 93-97.

Bainbridge, A. E., and O'Brien, B. J., 1962: Levels of tritium in a variety of New Zealand waters and some tentative conclusions from these results (in Tritium in the Physical and Biological Sciences). Vienna, IAEA, p. 33-39.

Begemann, F., 1963: New measurements on the worldwide distribution of naturally and artificially produced tritium (in Peaceful Uses of Atomic Energy, 2nd International Conference, 1963, Geneva). Fed. Rep. 18, p. 545.

Brown, R. M., 1970: Distribution of hydrogen isotopes in Canadian waters (in Isotope Hydrology, Symposium Proceedings, 1970, Vienna). Vienna, IAEA, STI/PUB/255, paper n. SM-129/1, p. 3-21.

Cambray, R. S., Fisher, E. M. R., Brooks, W. L., and Peirson, D. H., 1970: Radioactive fallout in air and rain--results to the middle 1970s. Atomic Energy Research Establishment, Harwell, England, Rpt. AERE-R-6556 (NTIS), 53 p.

Carmi, I., and Gat, J. R., 1973: Tritium in precipitation and fresh water sources in Israel. Israel J. Earth Sci. v. 22 n. 2, p. 71-92.

Dincer, T., and Payne, B. R., 1971: An environmental isotope study of the southwestern karst region of Turkey. J. of Hydrol., v. 14, n. 3-4, p. 233-258.

Drobinski, J. C., Jr. et al., 1965: Analyses of environmental samples for carbon-14 and tritium. Health Physics v. 11, n. 5, p. 385-395.

Eberline Instrument Corp., 1972: Environmental monitoring report--period covering May 1, 1972-July 31, 1972. Santa Fe, New Mexico, Dept. of Nuclear Sciences, 22 p.

Environmental Protection Agency, 1972: Tritium surveillance system, January-March 1972. Radiation Data and Rpts., v. 13, p. 438-441.

Fireman, E. L., and Schwarzer, D., 1954: Measurement of the tritium concentration in natural waters by a diffusion cloud chamber. Phys. Rev., v. 94, n. 2, p. 385-388.

Foster, R. F., 1973: Sources and inventory of radioactivity in the aquatic environment. Battelle Pacific Northwest Lab. Rpt. BNWL-SA-4614 (NTIS), 26 p.

Hughes, M. W., and Dighton, J. C., 1972: Tritium concentration of a variety of water samples. Commonwealth Scientific and Industrial Research Organization (Glen Osmond, Australia), Div. of Soils, Tech. Memorandum n. 4 (NTIS), 6 p.

International Atomic Energy Agency, 1970: Environmental isotope data, world survey of isotope concentration in precipitation (1964-1965). Vienna, IAEA, Tech. Rpt. 2 Ser. n. 117.

International Atomic Energy Agency, 1971: Environmental isotope data, no. 3, world survey of isotope concentration in precipitation (1966-1967). Vienna, IAEA, Tech. Rpt., Ser. n. 129, 402 p.

International Atomic Energy Agency, 1973: Environmental isotope data, no. 4, world survey of isotope concentration in precipitation (1969-1969). Vienna, IAEA, Tech. Rpt. Ser. n. 147, 334 p.

Institute of Nuclear Sciences, 1970: Progress report (no. 15), Jan.-June, 1970. Rpt. n. INS-458 (NTIS), 80 p.

Leventhal, J. S., and Libby, W. F., 1968: Tritium geophysics from 1961-1962 nuclear tests. J. Geophys. Res., v. 73, n. 8, p. 2715-2719.

Leventhal, J. S., and Libby, W. F., 1970: Tritium fallout in the Pacific United States. J. Geophys. Res., v. 75, n. 36, p. 7628-7633.

Libby, W. F., 1954: Datierung mittels radioaktiven kohlenstoffs und tritiums. Möglichkeit industrieller verwendung dieser isotope (Dating by means of radioactive carbon and tritium. Possibility of industrial application of these isotopes). Zeitschr. Elektrochemie, Band 58, n. 8, p. 574-585.

Libby, W. F., 1961: Tritium geophysics. J. Geophys. Res. v. 66, n. 11, p. 3767-3784.

Libby, W. F., 1962: Tritium geophysics, recent data and results (in Symposium on the Detection of Tritium in the Physical and Biological Sciences, 1961, summaries in French, Russian and Spanish). Vienna, IAEA, v. 1, p. 5-32.

Mullins, J. W., and Stein, J. L., 1972: Evaluation of tritium in ground and surface water of the western United States, April 1968-December 1969. Radiation Data and Rpts., v. 13, n. 2, p. 59-67.

National Institute of Radiological Sciences, 1972: Radioactivity survey data in Japan. Rpt. n. NIRS-RSD-34 (NTIS), 25 p.

National Environmental Research Center, 1973: Tritium surveillance system, April-June 1973. Radiation Data and Rpts., v. 14, n. 10, p. 601-604.

National Environmental Research Center, 1974: Off-site surveillance activities of the national environmental research center from July through December 1970. NERC-LV-539,17 (NTIS), 153 p.

Office of Radiation Programs, 1973: Radiation data, section II, water. Radiation Data and Rpts., v. 14, n. 2, p. 102-114.

Office of Radiation Programs, 1973: Tritium surveillance system, October-December 1972. Radiation Data and Rpts. v. 14, n. 5, p. 294-297.

Ostlund, H. G., and Lundgren, L. B., 1964: Stockholm natural tritium measurements. Tellus, v. 16, n. 1, p. 118-131.

Suess, H. E., 1964: Determination of tritium in natural waters. NTIS, PB-174 123, 2 p.

Takahashi, T., Ohno, S., and Hamada, T., 1963: Measurement of low-level tritium. 5th Japan Conf. Radioisotopes, Proc. n. 3, p. 75-77.

Thatcher, L. L., and Hoffman, C. M., 1963: Tritium fall-out over North America from the Soviet tests in 1961. J. Geophys. Res., v. 68, n. 20, p. 5899-5901.

Vinogradov, A. P., Devirts, A. L., and Dobkina, E. I., 1968: The current tritium contents of natural waters. Geochemistry International, v. 5, n. 5, p. 952-966.

B. INTEGRATED AREAL STUDIES

Anderson, L. J., 1966: Tritium ind holdet i grundvandet og dets betydning ved geohydrologiske undersøgelser (with English summary). Dan. Geol. Foren., Medd., v. 16, pt. 2, p. 177-190.

Langstrvp area and Hjondkaer, Denmark

Arnason, B., and Sigurgeirsson, Th., 1967: Hydrogen isotopes in hydrological studies in Iceland (in Isotopes in Hydrology, Symposium of International Atomic Energy Agency and International Union Geodesy and Geophysics, November 14-18, 1966, Vienna). Vienna, IAEA, STI/PUB/141, paper n. SM 83/3, p. 35-47.

Includes steam, deuterium

Atomic Energy of Canada, Ltd., 1970: The Perch Lake evaporation study--the use of radioisotopes in hydrology, February 1970, Vienna. Atomic Energy of Canada, Ltd. Rpt. AECL-3557.

Bolin, B., 1962: An investigation of tritium in atmospheric moisture, rain water and the sea in the European area. Progress Rpt., July 1, 1961-June 30-1962. U.S. Atomic Energy Comm., TID-16272, 14 p.

Sweden, Norway, Iceland and Ireland, 1957-1962

Brown, R. M., 1961: Hydrology of tritium in the Ottawa Valley. Geochim. Cosmochim. Acta, v. 21, n. 3-4, p. 199-215.

Carlston, C. W., 1964: Tritium-hydrologic research, some results of the U.S. Geological Survey research program. Science, v. 143, p. 804-806.

Small basins in Wisconsin and New Jersey

Eriksson, E., 1970: Investigation of tritium in atmospheric moisture, rain water, ground water and the sea in the European area. Univ. Stockholm Institute of Meteorology, Rpt. AC-9, 7 p.

Eriksson, E., 1971: Compartment models and reservoir theory (in Annual Review of Ecology and Systematics, v. 2). Palo Alto, California, Annual Reviews Inc., p. 67-84.

Florkowski, T., 1974: Ergebnisse des einsatzes von radioisotopen in der hydrogeologie der VR Polen (Results obtained with hydrogeologic uses of radioisotopes in Poland), English summary. Z. Angew. Geol., v. 20, n. 3, p. 107-111.

Fontes, J.-C., and Létolle, R., Olive, Ph., and Blavoux, B., 1967: Oxygen-18 and tritium in the Evian Basin (in Isotopes in Hydrology, Symposium of International Atomic Energy Agency and International Union Geodesy and Geophysics, November 14-18, 1966, Vienna). Vienna IAEA, STI/PUB/141, paper n. SM-83/28, p. 401-415.

Grant-Taylor, T. L., and Taylor, C. B., 1967: Tritium hydrology in New Zealand (in Isotopes in Hydrology, Symposium of International Atomic Energy Agency and International Union Geodesy and Geophysics, November 14-18, 1966 Vienna). Vienna, IAEA, STI/PUB/141, paper n. SM 83/27, p. 381-400.

Hufen, T. H., Duce, R. A., and Lau, L. S., 1969: Some measurements of the tritium content in the natural waters of southern Oahu, Hawaii. Hawaii Water Resources Research Center, Tech. Rpt. 34 (NTIS), 22 p.

Hufen, T. H., Buddemeier, R. W., and Lau, L. S., 1972: Tritium and radiocarbon in Hawaiian natural waters, Part 1. Hawaii Water Resources Research Center, Completion Rpt. 53 (NTIS), 54 p.

Isotopes, Inc., 1965: Technical report--tritium analyses of surface, spring and well water from the San Antonio area, Texas (in Survey Report of Edwards Underground Reservoir, Guadalupe, San Antonio and Nueces Rivers and Tributaries, Texas, v. 3, app. 3). Fort Worth, Texas, U.S. Army Corps Engineers, 41 p.

Jordan, C. F., and Drewry, G. E., 1969: The rain forest project, annual report. U.S. Atomic Energy Comm., Depository Libraries, AEC Tech. Rpt. PRNC-129, 144 p.

Kraynova, L. P., 1964: Kizucheniyu izotopnogo sostava vod vysokogornykh istochnikov (On the study of the isotope composition of waters from high mountain sources). Akad. Nauk Uzbek. SSR Uzbek. Geol. Zhur., n. 6, p. 83-85.

Lau, L. S., and Hailu, T., 1968: Tritium measurements in natural Hawaiian waters, instrumentation. Hawaii Water Resources Research Center, Tech. Rpt. n. 22, 36 p.

Summarizes all known measurements

Lau, L. S., 1973: Tritium measurement of natural waters on Oahu, Hawaii, a preliminary interpretation (sampling period-July 1969 to June 1970). Hawaii Water Resources Research Center, Tech. Rpt. n. 65 (NTIS), 39 p.

Lévèque, P., and Vigneaux, M., 1965: Results of measurements of tritium activity in certain waters of the southwest. Acad. Sci. Comptes Rendus, v. 261, n. 20, p. 4168-4171.

Lévèque, P., Vigneaux, M., and Alvinerie, J., 1967: Evaluation of tritium activity in the southwest of France (in Isotopes in Hydrology, Symposium of International Atomic Energy Agency and International Union Geodesy and Geophysics, November 14-18, 1966, Vienna). Vienna, IAEA paper n. SM-83/30, p. 417-432.

Lévèque, P., Gros, J. C., Maurin, C., et al., 1972: Utilization du tritium et des télédétections γ et infrarouge dans l'étude hydrologique de la région de Xanthe-Komotini en Thrace orientale (Grece septentrionale) (The use of tritium and gamma ray and infrared remote sensing in the hydrologic study of the Xanthe-Komotine region of eastern Thrace, northern Greece). Acad. Sci. Comptes Rendus, Ser. D., v. 274, n. 17, p. 2447-2450.

Leventhal, J. S., and Libby, W. S., 1973: Tritium hydrology studies in California (in Tritium). Phoenix, Arizona, Messenger Graphics, p. 724-736.

Mangano, F., Marce, A., Meybeck, M., et al., 1970: Idrologia isotopica metodologia e prime applicazioni alle sorgenti Fiume e Bella, Madonie orientali (Isotope hydrology, methodology and primary application to the sources at Fiume and Bella, eastern Madonie Mountains, French summary). Riv. Min. Sicil., v. 21, n. 124-126, p. 223-231.

Mangano, F., Marce, A., Meybeck, M., et al., 1971: Idrologia isotopica metodologia e prime applicazioni alle sorgenti Fiume e Belle-Madonti orientali (Isotopic hydrology, methodology and initial application to the springs of Fiume and Bella, eastern Madonie Mountains) (in Convegno Internazionale sulle acque Sotterranee, Atti, French summary). Palermo, Ente Sviluppo Agric. Sicil., Assoc. Int. Idrogeol., p. 609-616.

Mitchell, N. T., 1971: Radioactivity in surface and coastal waters of the British Isles, 1970. Lowestoft, England, Ministry of Agriculture, Fisheries and Food, Fisheries Radiobiological Lab. (NTIS), 28 p.

Morgante, S., Mosetti, F., and Tongiorgi, E., 1966: Moderne indagini idrologiche nella zona di Gorizia (Recent hydrological investigations in the Gorizia zone, English summary). Boll. Geofisica Teor. ed Appl., v. 8, n. 30, p. 114-137.

Plata, A., Sanchez, W., and Szulak, C., 1973: Studies of aquifers of the Gurgueia and Fidalgo River Basins in the state of Piaui using environmental isotopes (in Portuguese). Sao Paulo, Brazil, Instituto de Energia Atomica (NTIS), 46 p.

Rabinowitz, D., and Gross, G. W., 1972: Environmental tritium as a hydrometeorological tool in the Roswell Basin, New Mexico. Las Cruces, New Mexico, Water Resources Research Institute, Rpt. 016, 268 p.

Rabinowitz, D., and Gross, G. W., 1973: Precipitation/recharge relation for a limestone aquifer obtained by natural tritium studies (abstr.). Am. Geophys. Union, Trans., Eos, v. 54, n. 4, p. 262.

Roether, W., 1966: Estimating the tritium input to ground water from wine samples; ground water and direct runoff contribution to central European surface water (in Isotopes in Hydrology, Symposium of International Atomic Energy Agency and International Union Geodesy and Geophysics, November 14-18, 1966, Vienna). Vienna, IAEA STI/PUB/141, paper n. SM-83/7, p. 73-91.

Siegenthaler, U., Oeschger, and Tongiorgi, E., 1970: Tritium and oxygen-18 in natural water samples from Switzerland (in Isotope Hydrology, 1970, Symposium of International Atomic Energy Agency and UNESCO, March 9-13, 1970, Vienna). Vienna, IAEA, STI/PUB/255, paper n. SM-129/22, p. 373-385.

Smith, D. B., 1973: Tritium water tracing. Water Treatment and Examination, v. 22, n. 4, p. 250-258.

Thatcher, L., Burin, M., and Brown, G. F., 1961: Dating desert ground water. Science,v. 134, n. 3472, p. 105-106.

Theodorsson, P., 1967: Natural tritium in ground water studies (in Isotopes in Hydrology, Symposium of International Atomic Energy Agency and International Union Geodesy and Geophysics, November 14-18, 1966, Vienna). Vienna, IAEA, STI/PUB/141, paper n. SM-83/126, p. 371-380.

Theodorsson, P., 1968: Thríventi í grunnvanti og jöklum á Islandi (Tritium in ground water and glaciers in Iceland, with English summary and captions). Jökull, v. 18, p. 350-358.

Verhagen, B. Th., Sellschop, J. P. F., and Jennings, C.M.H., 1970: Contribution of environmental tritium measurements to some geohydrological problems in Southern Africa (in Isotope Hydrology, 1970 Symposium of International Atomic Energy Agency and UNESCO, March 9-13, 1970, Vienna). Vienna, IAEA, STI/PUB/255, paper n. SM-129/58, p. 283-313.

Verhagen, B. Th., 1971: Tritium in hydrology. S. African J. Sci., v. 67, n. 4, p. 289-295.

Vogel, J. C., Ehhalt, D., and Roether, W., 1963: A survey of the natural isotopes of water in South Africa (in Radioisotopes in Hydrology, with French, Russian and Spanish abstracts). Vienna, IAEA, p. 407-415.

Von Buttlar, H., 1963: Stratosphärisches tritium (Stratospheric tritium). Geochim. Cosmochim. Acta, v. 27, n. 7, p. 741-751.

C. ATMOSPHERE

Bainbridge, A. E., Suess, H. E., and Friedman, I., 1961: Isotopic composition of atmospheric hydrogen and methane. Nature, v. 192, n. 4803, p. 648-649.

1948-1960 survey

Bainbridge, A. E., 1967: Comments on E. Eriksson's account of the major pulses of tritium and their effects in the atmosphere. Tellus, v. 19, n. 4, p. 645.

Bard, S. T., 1974: Entrainment of stratospheric tritium, another radiological anomaly of the Rocky Mountain West (in 8th Midyear Topical Symposium of Health Physics Soc., October 21-24, 1974, Knoxville, Tennessee). CONF. 741018, p. 73-79.

Begemann, F., and Friedman, I., 1959: Tritium and deuterium content of atmospheric hydrogen. Zeitschr. Naturforsch., v. 14a, n. 12, p. 1024-1031.

1954-1956 Buffalo, New York; Hamburg and Nuremberg, Germany

Begemann, F., 1961: Tritium determinations in atmospheric gases and meteorites. Mainz Max-Planck-Institut f. Chemie (Otto-Hahn Institut), Contract AF-61(052)-465, Summary Rpt. n. 1, 6 p.

November 1960 - October 1961 survey

Begemann, F., 1963: Tritium content of atmospheric hydrogen and atmospheric methane (in Earth Science and Meteoritics). Amsterdam, North-Holland Pub. Co., p. 167-187.

Dates to 1962, Germany

Begemann, F., 1963: Tritium content of atmospheric hydrogen and atmospheric methane. J. Geophys. Res., v. 68, n. 13, p. 3757-3758.

Summary of experimental studies

Begemann, F., and Friedman, I., 1968: Tritium and deuterium in atmospheric methane. J. Geophys. Res., v. 73, n. 4, p. 1149-1153.

Gary, Indiana, and Oberhausen, Germany

Begemann, F., and Friedman, I., 1968: Isotopic composition of atmospheric hydrogen. J. Geophys. Res., v. 73, n. 4, p. 1137-1147.

Oberhausen and München, Germany

Beran, M., and Assaf, G., 1970: Use of isotopic decay rates in turbulent dispersion studies. J. Geophys. Res. v. 75, n. 27, p. 5279-5285.

Bibron, R., 1965: Détection du tritium atmosphérique par scintillation; évolution de sa concentration en France (Detection of atmospheric tritium by scintillation; variations on its concentrations in France). Univ. Paris Ph.D. thesis, 104 p.

Bierly, E. W., 1967: Application of isotopes to some problems in atmospheric sciences (in Symposium on Isotope Techniques in the Hydrologic Cycle, November 1965, Univ. Illinois). Am. Geophys. Union, Geophysical Mon. Ser. n. 11, p. 37-46.

Review papers

Bishop, K. F., Delafield, H. J., Eggleton, A. E. J., et al., 1962: Tritium content of atmospheric methane (in Symposium on the Detection and Use of Tritium in the Physical and Biological Sciences, 1961, Vienna). Vienna, IAEA, v. 1, p. 55-67.

Variations with altitude, Northern and Southern Hemispheres

Bolin, B., 1959: The use of tritium in the study of vertical exchange in the atmosphere (summary only). Adv. Geophys., v. 6, p. 297-298.

Sweden study

Bradley, W., and Stout, G. E., 1970: Vertical distribution of tritium in water vapor in the lower troposphere. Tellus, v. 22, n. 6, p. 699-706.

Brown, F., Goldsmith, P., Green, H. F., Holt, A., and Parham, A. G., 1961: Measurements of the water vapour, tritium and carbon-14 content of the middle stratosphere over southern England. Tellus, v. 13, n. 3, p. 407-416.

Carbon-14 and deuterium, 80,000-100,000 feet

Cambray, R. S., Fisher, E. M., Brooks, W. L., and Pierson, D. H., 1969: Radioactive fallout in air and rain; results to the middle of 1969. London, H. M. Stationery Office, U.K. Atomic Energy Auth. Res. Gr. Rpt. AERE-R6216, 55 p.

January 1966 to Mid 1969, three stations in the United Kingdom

Ehhalt, D., Israel, G., Roether, W., Stich, W., et al., 1963: Tritium and deuterium content of atmospheric hydrogen. J. Geophys. Res., v. 68, n. 13, p. 3747-3751.

Ehhalt, D., 1966: Tritium and deuterium in atmospheric hydrogen (with Russian and English summaries). Tellus, v. 18, n. 2-3, p. 249-255.

Ehhalt, D. H., 1971: Vertical profiles and transport of HTO in the troposphere. J. Geophys. Res., v. 76, n. 30, p. 7351-7367.

1965 to 1967, 0 to 9.2 km; Scottsbluff, Nebraska; 1966 to 1967, Eastern Pacific and Death Valley

Ehhalt, D. H., 1973: Turnover times of ^{137}Cs and HTO in the troposphere and removal rates of natural aerosol particles and water vapor. J. Geophys. Res., v. 78, n. 30, p. 7076-7086.

Eriksson, E., 1965: Account of the major pulses of tritium and their effects in the atmosphere. Tellus, v. 17, n. 1, p. 118-130.

Facy, L., 1958: Quelques idées exprimées du point de vue strictement météorologique (Some ideas expressed from a strictly meteorological point of view) (in International Geophysical Year, 1957/1958, Annals, 5). IGY Instruction Manual, Pt. 5, p. 340-343.

Fireman, E. L., and Rowland, F. S., 1961: An additional measurement of the tritium content of atmospheric hydrogen of 1949. J. Geophys. Res., v. 66, n. 12, p. 4321.

Frejaville, G., 1967: Atmospheric tritium measurement and application. Centre d'Etudes Nucléaires, Rpt. CEA-76 (NTIS), 24 p.

Friedman, I., 1974: Isotopic composition of atmospheric hydrogen, 1967-1969. J. Geophys. Res., v. 79, n. 6, p. 785-788.

Mount Erane, Colorado

Goldsmith, P., Jelley, J. V., Barclay, F. R., Elliot, M. J., and Osborne, A. R., 1960: Some preliminary measurements of the tritium and carbon-14 content of the stratosphere over England. Harwell, England, Atomic Energy Research Establishment, Rpt. AERE-R3271, 9 p.

90,000 feet

Gonsier, B., and Friedman, I., 1962: Tritium and deuterium in atmosphärischen wässerstoff (Tritium and deuterium in atmospheric hydrogen). Zeitschr. Naturforsch., v. 17a, n. 12, p. 1088-1091.

Gonsier, B., Friedman, I., and Ehhalt, D., 1963: Measurements of tritium and deuterium concentration in atmospheric hydrogen. J. Geophys. Res., v. 68, n. 13, p. 3753-3756.

1957 to 1958, closely spaced samples

Gonsier, B., Friedman, I., and Lindenmayr, G., 1966: New tritium and deuterium measurements in atmospheric hydrogen. Tellus, v. 18, n. 2-3, p. 251-261.

Goodman, P., 1974: Tritium exchange detection of stratospheric water vapor concentration. NTIS, AD/A-001, 336/75L, 39 p.

Haines, A., and Musgrave, B. C., 1968: Tritium content of atmospheric methane and ethane. J. Geophys. Res., v. 73, n. 4, p. 1167-1173.

Neosho, Missouri study

Junge, C. E., 1964: Neuere ergebnisse de luftchemie und ihre bedeutung für die meteorologie (Recent results in atmospheric chemistry and their meteorological significance). Deutscher Wetterdienst, Berichte, v. 12, n. 91, p. 80-88.

Junge, C. E., 1963: Studies of global exchange processes in the atmosphere by natural and artificial tracer. J. Geophys. Res., v. 68, n. 13, p. 3849-3856.

Kigoshi, K., and Yoneda, K., 1974: Daily variations in the tritium concentration of atmospheric moisture. J. Geophys. Res., v. 75, n. 15, p. 2981-2984.

1966, Tokyo

Krey, P. W., and Krajewski, B., 1970: Troposphere scavenging of ^{90}Sr and ^{3}H (<u>in</u> Proceedings of Symposium, Richland, Washington, June 2-4, 1970). Conf. n. 700601, AEC Symposium Series-22 (NTIS), p. 447-463.

Kulp, J. L., Bruno, J., and Broecker, W. S., 1956: Radioactive isotopes in the atmosphere. Columbia Univ. Lamont Geological Laboratory, Contract AF 19(604)-851, 41 p.

Carbon-14 and deuterium, 401 samples

Layton, B. R., 1969: Tritium in atmospheric hydrogen and atmospheric water. Arkansas Univ., Thesis (NTIS) 62 p.

1963 to 1966, correlation of T/H with D/H

Libby, W. F., 1958: Isotopes in meteorology. Am. Meteor. Soc. Bull., v. 39, n. 2, p. 65-68.

McKay, L. R., and Witherspoon, J. P., 1974: Effect of stack height of individual and population doses attributable to gaseous effluents from a model fuel reprocessing facility (in 8th Midyear Topical Symposium of the Health Physics Society, October 21-24, Knoxville, Tennessee). CONF 741018, p. 223-228.

Machta, L., 1958: Some applications of radioactive tracers to large-scale meteorological problems (in International Geophysical Year 1957/1958 Annals). IGY Instruction Manual, Pt. 5, p. 313-323.

Radon

Machta, L., 1960: Radioactive tracers in hydrology. Am. Soc. Civil Engineers, Hydraulics Div., J. Hy., v. 4, p. 49-60.

Radon

Malurkar, S. L., 1956: Radioactive isotopes for study of tropical meteorology and oceanography. J. Sci. Industrial Research, Sec. A., v. 15, n. 6, p. 272-274.

Martell, E. A., 1963: On the inventory of artificial tritium and its occurrence in atmospheric methane. J. Geophys. Res., v. 68, n. 13, p. 3759-3769.

Martin, D., and Hackett, J. P., 1974: Tritium in atmospheric hydrogen. Tellus, v. 26, n. 5, p. 603-608.

1971 to 1973, West Wood, New Jersey

Miyake, Y., 1958: Radioactivity as a tracer of air motions in the atmosphere (in International Geophysical Year 1957/1958 Annals). IGY Instruction Manual, Pt. 5, p. 360.

Münnich, K. O., and Roether, W., 1967: Transfer of bomb ^{14}C and tritium from the atmosphere to the ocean; internal mixing of the ocean on the basis of tritium and ^{14}C profiles (in Symposium on Radioactive Dating and methods of low-level counting, March 2-10, 1967, Monaco). Vienna, IAEA, p. 93-104.

Ostlund, H. G., 1967: Hurricane tritium I; preliminary results on Hilda 1964 and Betsy 1965. Am. Geophys. Union Geophysical Mon. Ser., n. 11, p. 58-60.

Ostlund, H. G., 1971: Atmospheric HT and HTO 1970-71, Tech. Rpt., May 15, 1970-August 14, 1971. NTIS, ORO-3944-3, 26 p.

Miami and Fairbanks

Ostlund, H. G., 1973: Tritium in the atmosphere and oceans (in Tritium). Phoenix, Arizona, Messenger Graphics, p. 382-391.

Ostlund, H. G., 1968: Tritium in hurricanes. NTIS, PB-182-485, 9 p.

Schell, W. R., Sauzey, G., and Payne, B. R., 1970: Tritium injection and concentration distribution in the atmosphere. J. Geophys. Res., v. 75, n. 12, p. 2251-2266.

Scholz, T. G., 1970: Water vapor, molecular hydrogen, methane, and tritium concentrations near the stratosphere. J. Geophys. Res., v. 75, n. 12, p. 3049-3054.

Smith, R. M., 1968: A preview of the determination of mass return flow of air and water vapor into the stratosphere using tritium as a tracer. Tellus, v. 20, n. 1, p. 76-81.

July 1962, North America

Taylor, G. B., 1971: Influence of 1968 French thermonuclear tests on tritium fallout in the Southern Hemisphere. Earth and Planet. Sci. Lett., v. 10, n. 2, p. 196-198.

1969 to 1970, Southwest Pacific

Wolfgang, R., 1961: Origin of high tritium content of atmospheric methane, hydrogen and stratospheric water. Nature, v. 192, n. 4809, p. 1279-1280.

Yoneda, K., and Kigoshi, K., 1968: Atmospheric tritium as an indicator of air masses (Japanese and English summaries). Chikuykagaku (Geochem), v. 2, n. 1, p. 1-7.

D. PRECIPITATION

Allison, G. B., Holmes, J. W., and Hughes, M. H., 1971: Tritium fallout in Southern Australia and its hydrologic implication. J. Hydrol., v. 14, n. 3/4, p. 307-321.

Begemann, F., 1959: Neubestimmung der natürlichen irdisher tritiumzerfallsrate und die frage der herkunft des "natürlichen" tritium (New determination of the natural terrestrial decay rate of tritium and the question of the origin of "natural" tritium). Zeitschr. Naturforsch., v. 14, n. 4, p. 334-342.

Bibron, R., Delibrias, G., and Labeyries, J., 1963: Evolution de la concentration du tritium dan les eaux de précipitation en France (Evolution of the tritium concentration in the rainwater in France). Paris, Acad. Sci. Comptes Rendus, v. 256, n. 23, p. 4951-4954.

Booker, D. V., 1965: Exchange between water droplets and tritiated water vapor. Royal Meteorolog. Soc., Quarterly Jour., v. 91, n. 387, p. 73-79.

Brown, R. M., and Grummitt, W. E., 1956: The determination of tritium in natural waters. Canadian J. Chem., v. 34, n. 3, p. 220-226.

1951 to 1956, Ottawa

Dincer, T., Martinec, J., Payne, B. R., and Yen, C. K., 1970: Variation of the tritium and oxygen-18 content in precipitation and snowpack in a representative basin in Czechoslovakia (<u>in</u> Isotope Hydrology 1970, Symposium of International Atomic Energy Agency and UNESCO, March 9-13, 1970, Vienna). Vienna, IAEA, ST1/PUB/255, Paper n. SM-129/3, p. 23-42.

1965 to 1966, oxygen-18, snow and snowpack

Eriksson, E., et al., 1970: Tritium data on precipitation and river water in the Nordic countries. Stockholm Univ. Institute Meteorol. Rpt. AC-6, 7 p.

Friedman, I., Machta, L., and Soller, R., 1962: Water vapor exchange between a water droplet and its environment. J. Geophys. Res., v. 67, n. 7, p. 2761-2766.

Deuterium

Gat, J. R., Karfunkel, U., and Nir, A., 1961: Tritium content of rain water from the eastern Mediterranean area. U.S. Atomic Energy Comm., TID-12612, 28 p.

1958 to 1961, Israel and eastern Mediterranean countries

Gat, J. R., and Carmi, I., 1970: Evolution of the isotopic composition of atmospheric waters in the Mediterranean Sea area. J. Geophys. Res., v. 75, n. 15, p. 3039-3048.

Deuterium and oxygen-18; 20 coastal/island stations

Gonsier, B., 1959: Tritium-anstieg im atmosphärischen wasserstoff (Increase of tritium in atmospheric hydrogen). Die Naturwissenschaften, v. 46, n. 6, p. 201-202.

Gulliksen, S., 1970: Tritium content of rain in Trondheim. J. Geophys. Res., v. 75, n. 12, p. 2247-2249.

1967, Norway

Hales, J. M., and Schwendiman, L. C., 1971: Precipitation scavenging of tritium and tritiated water. Richland, Washington, Battelle Memorial Institute, Pacific Northwest Lab., Annual Rpt. 1970, v. 2, Physical Sciences, pt. 1, Atmospheric Sciences, p. 82-84.

Methods

Hales, J. M., 1972: Scavenging of gaseous tritium compounds by rain. Battelle Pacific Northwest Lab. Rpt. BNWL-1659 (NTIS), 35 p.

Theory

Israel, G. W., 1962: Messung des tritium-jahresganges im regen 1960-1961 nach isotopenanreicherung im trennrohr (Measurement of the annual course of tritium in rain in 1960-1961 according to isotopic enrichment in the separating tube) English abstr. Zeitschr. Naturforsch., v. 17a, n. 10, p. 925-929.

Cologne

Israel, G., Roether, W., and Schumann, G., 1963: Seasonal variations of bomb-produced tritium in rain. J. Geophys. Res., v. 68, p. 3771-3773.

1960 to 1961, Sindorf, Germany

Libby, W. F., 1954: Rain three weeks old. Sci. News Lett., v. 66, n. 13, p. 197.

Libby, W. F., 1963: Moratorium tritium geophysics. J. Geophys. Res., v. 68, n. 15, p. 4485-4494.

1958, Sr-90

McNaughton, D. L., 1970: Relationship between the trajectory of air parcels reaching Rhodesia and the tritium content of the rain (in Symposium on Tropical Meteorology, Univ. Hawaii, June 2-11, 1970, Honolulu). Am. Meteorolog. Soc., Sec. B-IX, 4 p.

Moghissi, A. A., and Porter, C. P., 1970: Tritium concentration in precipitation, 1967-1968. Radiological Health Data and Rpts., v. 11, n. 3, p. 137-140.

10 states in the United States including Alaska and Hawaii

Payne, B. R., and Thatcher, L. L., 1967: The occurrence of tritium in precipitation and its significance in hydrological investigations (in Proceedings International Conference on Water for Peace, Washington, D.C., V. 4). Vienna, IAEA, p. 320-331.

Deuterium, oxygen-13; 100 stations throughout the world

Soyner, V. N., et al., 1971: Ob ekraniruyushchem effekte kontinentov opredelyayushchem izotopnyy sostav atmosfernoy vlagi, po dannym prostranstvenno-vremennogo raspredeleniya tritiya v atmosfernov asadkakh umerennykh shirot severnogo polushariya (Screening effect of continents determining the atmospheric moisture isotopic composition, based on data concerning tritium space time distribution in atmospheric precipitation on the Northern Hemisphere temperate latitudes). Akademiya Nauk SSSR, Doklady, Ser. Matem., Fizika, v. 201, n. 1-3, p. 78-81.

Stewart, G. L., and Hoffman, C. M., 1966: Tritium rainout over the United States in 1962 and 1963. U.S. Geol. Survey Circ. 520, 11 p.

1962 to 1963, United States and Puerto Rico, 15 stations

Stewart, G. L., and Farnsworth, R. K., 1968: United States tritium rainout and its hydrologic implication. Water Resources Res., v. 4, n. 2, p. 273-289.

Stewart, G. L., and Wyerman, T. A., 1970: Tritium rainout in the United States during 1966, 1967 and 1968. Water Resources Res., v. 6, n. 1, p. 77-87.

Stout, G. E., and Huff, F. A., 1967: Rainout characteristics for hydrological studies (in Isotopes in Hydrology Symposium of International Atomic Energy Agency and International Union Geodesy and Geophysics, November 14-18, 1966, Vienna). Vienna, IAEA, pp. 61-72.
Variation over 10-12 square miles, ± 20-25 percent

Von Buttlar, H., Stahl, W., and Wiik, B., 1962: Tritiummessungen an regenwasser ohne isotopenan reichering (Tritium measurements on rainwater without isotope enrichment). Zeitschr. Naturforsch., v. 17a, n. 1, p. 91-92.

Von Buttlar, H., 1963: Tritium in rainwater (in Earth Science and Meteoritics). Amsterdam, North-Holland Publishing Co., p. 188-206.

E. EVAPORATION AND TRANSPIRATION

Allison, G. B., 1974: Estimation of ground water accession to and evaporation from a South Australian lake using environment tritium. Australian J. Soil Res., v. 12, n. 2, p. 119-131.

Barry, P. J., 1967: The use of radioactive tracer gases to study the rate of exchange of water vapor between air and natural surfaces (in Isotope Techniques in the Hydrologic Cycle). Am. Geophys. Union, Geophysical Mon. Ser., n. 11, p. 69-76.

Barry, P. J., 1970: Perch Lake evaporation study (in Isotope Hydrology, 1970 Symposium of International Atomic Energy Agency and UNESCO, March 9-13, 1970, Vienna). Vienna, IAEA, STI/PUB/255, paper n. SM-129/10, p. 139-151.

Horton, J. H., Corey, J. C., and Wallace, R. M., 1971: Tritium loss from water exposed to the atmosphere. Environmental Sci. Technol., v. 5, n. 4, p. 338-343.

Machta, L., 1969: Evaporation rates based on tritium measurements for hurricane Betsy. Tellus, v. 21, n. 3, p. 404-408.

Ostlund, H. G., 1970: Modification of atmospheric tritium and water vapor by Lake Tahoe. Tellus, v. 22, n. 4, p. 463-465.

Van der Westhuizen, M., and Smith, M. J., 1971: Change in concentration of tritium in water during evaporation and the possibility of using it to determine evaporation from water,soil and plant surfaces. South African Atomic Energy Board, PEL-214 (NTIS), 15 p.

F. RIVERS AND SURFACE WATER HYDROLOGY

Barnes, J., and Carswell, D., 1972: Tritium measurements of the Murrumbidgee River at Wagga. Geol. Soc. Aust., J., v. 19, pt. 2, p. 225-227.

Australia

Chu, D. S. L., 1973: Measurement of flood flow and sediment transport by radioisotope method (<u>in</u> IAEA Research Contracts, 13th Annual Report, Radioisotopes and Radiation Applications). Vienna, IAEA, Tech. Rpt. Ser. n. 144, p. 175-176.

Clayton, C. G., 1963: A comparison of radioisotope methods for river flow measurement (<u>in</u> Radioisotopes in Hydrology, Symposium Proceedings, IAEA, March 5-9, 1963, Tokyo). Vienna, IAEA, p. 1-24 (discussion p. 58-61).

England study

Crouzet, E., Hubert, P., Olive, P., Siwertz, E.,and Marce, A., 1970: Tritium in investigation of surface hydrology, experimental determination of coefficient of runoff. NTIS, CSIRO-TRANS-10757, 5 p.

France study

Crouzet, E., Hubert, P., Olive, P., Siwertz, E., and Marce, A., 1970: Tritium in the measurement of surface hydrology. Experimental determination of the runoff coefficient. J. Hydrol., v. 11, p. 217-229.

Dincer, T., 1967: Application of radiotracer methods in streamflow measurements (<u>in</u> Isotopes in Hydrology, Symposium of International Atomic Energy Agency and International Union Geodesy and Geophysics, November 14-18, 1966, Vienna). Vienna, IAEA, p. 93-110.

Drobinski, J. C., Magno, P. J., and Goldin, A. S., 1968: Plutonium, tritium, and carbon-14 in man and the biosphere. 1st International Congress Radiat. Prot., Proc., p. 369.

Farvolden, R. N., 1971: Base-flow recession in Illinois. Illinois Univ., Water Resources Center, Rpt. WRC-RR-42 (NTIS), 57 p.

Florkowski, T., Davis, T. G., Wallander, B., and Prabhakar, R. D. L., 1969: The measurement of high discharges in turbulent rivers using tritium tracer. J. Hydrol., v. 8, n. 8, p. 249-264.

Tana River, Africa

Kwapinski, J., 1973: A characterization of radioactive contamination of the Vistula River (in Polish). Gaz Woda Tech Sanit., v. 47, n. 9, p. 302-305.

Mosetti, F., 1965: A new interpretation of a radioactive tracer experiment in the Timava River. Boll. Geofis. Teor. Appl., v. 7, p. 218-243.

Italy

Skausen, D. M., 1973: The analysis of tritium oxide from selected areas of the Connecticut River. Connecticut Univ., Inst. Water Resources, Completion Rpt. (NTIS), 9 p.

Tsivoglou, E. C., Cohen, J. B., and Shearer, S. D., 1968: Tracer measurement of stream reaeration; field studies. Water Pollution Control Fed. J., v. 40, n. 2, pt. 1, p. 285-305.

Jackson River, Virginia, and West Virginia

Von Buttlar, H., Farzine, K., and Wohlfahrt, H. D., 1965: Tritium concentration of German river waters measured with the proportional counting technique. Nucl. Instrum. Meth., v. 37, p. 288-292.

1961 to 1964, Rhine and Main

Wyerman, T. A., Farnsworth, R. K., and Stewart, G. L., 1970: Tritium in streams in the United States, 1961-1968. Radiological Health Data and Rpts., v. 11. n. 9, p. 421-439.

G. LAKES AND RESERVOIRS

Alvinerie, J., Dégot, B., Lévèque, P., and Vigneaux, M., 1966: Activité en tritium et caractéristiques chimiques des eaux du lac Pavin (Tritium activity and chemical characteristics of the waters of Lake Pavin). Acad. Sci. Comptes Rendus, Ser. D, v. 262, n. 8, p. 846-849.

Alvinerie, J., Dégot, B., Lévèque, P., and Vigneaux, M., 1966: Etude de l'activité en tritium du bassin versant du Mont-Cenis (Study of the tritium activity of the Mont-Cenis drainage basin). Acad. Sci. Comptes Rendus, Ser. D, v. 262, n. 1, p. 32-35.

Baranyi, S., 1973: The use of tritium measurement in the investigation of Lake Balaton (in Hydrology of Lakes Symposium, July 1973, Helsinki, Finland). Internat. Assoc. Hydrol. Sci. Publ. n. 109, p. 28-32.

Hungary

Bartsch, A. F., 1970: Accelerated eutrophication of lakes in the United States; ecological response to human activities. Environ. Pollution, v. 1, n. 2, p. 133-140.

Blavoux, B., and Olive, P., 1966: Premiers resultats sur la teneur en tritium des coux du lac Léman (First results on the tritium content of the waters of Lake Geneva). Acad. Sci. Comptes Rendus, Ser. D, v. 262, n. 24, p. 2445-2448.

Switzerland

Dincer, T., and Payne, B. R., 1967: An isotope study of lakes in the karst region of southern Turkey (in Hydrology of Fractured Rocks, v. 2, French summary). Internat. Assoc. Sci. Hydrol., Publ. n. 74, p. 654-661.

Elder, R. A., and Vigander, S., 1967: Ultra-low velocity measurement in a stratified reservoir by isotopic current meter (in Symposium on Isotope Techniques in the Hydrologic Cycle, Univ. Illinois, November 1965). Am. Geophys. Union, Geophysical Mon. Ser. n. 11, p. 81-84.

Frederick, B. J., 1963: Labeling lake water with tritium. Internat. J. Appl. Radiation Isotopes, v. 14, p. 401-404.

Lake McMillon, New Mexico

Gorham, E., and Hofstetter, R. H., 1971: Penetration of bog peats and lake sediments by tritium from atmospheric fallout. Ecology, v. 52, n. 5, p. 898-902.

Minnesota study

Hubert, P., Meybeck, M., and Olive, P., 1970: The use of tritium to study the water dynamics of Lake Geneva. C.r.hebd. Seanc. Acad. Sci., v. 270, Ser. D, p. 1298-1301.

Jones, L. M., and Ostlund, H. G., 1971: Carbon-14 age and tritium content of Lake Vanda, Wright Valley. Antarctic Jour., v. 6, n. 5, p. 200-201.

Lehman, A., 1971: Transport of radionuclides in sediments (in 3rd National Symposium on Radioecology, May 10-12, 1971, Oak Ridge, Tennessee, proc.). CONF-710501, p. 936-944.

McMahon, J. W., 1964: The dispersion of tritium in Perch Lake. Atom. Energy Canada Ltd., AECL-1889, 18 p.

Meybeck, M., Hubert, P., Martin, J. M., and Olive, P., 1970: Tritium study of the mixing of water in lake and estuaries with particular reference to the Lake of Geneva and the Gironde (in Isotope Hydrology 1970, Symposium of International Atomic Energy Agency and UNESCO, March 9-13, 1970, Vienna). Vienna, IAEA, STI/PUB/255, Paper n. SM-129/32, p. 523-541, 19 p.

France and Switzerland

Payne, B. R., 1970: Water balance of Lake Chala and its relation to ground water from tritium and stable isotope data. J. Hydrol., v. 11, n. 1, p. 47-58.

Kenya

Simpson, H. J., 1969: Crater Lake as a tritium integrator (abstr.). Eos, Amer. Geophys. Union, Trans., v. 50, n. 4, p. 142.

Simpson, H. J., 1970: Tritium in Crater Lake, Oregon. J. Geophys. Res., v. 75, n. 27, p. 5195-5207.

Soyfer, V. N., Brezgunov, V. S., Verbolov, V. I., Votintseu, K. K., and Romanov, V. V., 1970: Primeneniye izotopnogo metoda dlya izucheniya protsessov vodoobmenal ozera Baykal (Application of the isotopic method to study of mixing processes in Lake Baikal) (in Techeniya i diffuziya vod Baykala). Akademiya Nauk SSSR Sibirskoye Otdelemya, Limnologicheskiy Instit Trudy, v. 14, n. 34, p. 146-153.

Deuterium and oxygen-18, USSR

H. GLACIERS AND SNOW PACKS

Aegerter, S., Oeschger, H., Renaud, A., and Schmacher, E., 1969: Studies based on the tritium content of the samples (in Etudes physiques et chimiques sur la glace de l'Indlandsis du Groenland, 1959). Medd. Grönland, v. 177, n. 2, p. 76-88.

Ambach, W., Eisner, H., and Thatcher, L. L., 1967: Tritium content in the firn layers of an alpine glacier (in Comm. of Snow and Ice, Gen. Assembly of Bern, September-October 1967, proc.). Internat. Assoc. Sci. Hydrol. Publ. 79, p. 126-134.

Ambach, W., Eisner, H., and Url, M., 1973: Seasonal variations in the tritium activity of runoff from an alpine glacier, Kesselwandferer, Oetztal Alps, Austria (in Symposium on the Hydrology of Glaciers, September 7-13, 1969, Cambridge, England, proc.). Internat. Assoc. Sci. Hydrol. Publ. 95, p. 199-204.

Ambach, W., Eisner, H., and Sauzay, G., 1969: Tritium profiles in two firn cores from Alpine glaciers and tritium content in precipitation in the Alpine area. Archiv für Meteorlogie, Geophysik und Bioklimatologie, Ser. B, v. 17, n. 1, p. 93-104.

Ambach, W., and Dansgaard, W., 1970: Fallout and climate studies on firn cores from Carrefour, Greenland. Earth Planet. Sci. Lett., v. 8, n. 4, p. 311-316.

Arnason, B., 1969: The exchange of hydrogen isotopes between ice and water in temperature glaciers. Earth Planet. Sci. Lett., v. 6, n. 6, p. 423-430.

Deuterium, Iceland

Arnason, B., Buason, Th., Martinec, J., and Theodorsson, P., 1973: Movement of water through snow pack traced by deuterium and tritium (in The Role of Snow and Ice in Hydrology, Banff Symposium, September 1972). Internat. Assoc. Hydro. Sci. Publ. 107, v. 1, p. 299-312.

Bahadur, J., 1973: Some observations on the vertical fall-out profile on Grasubreen, a mountain glacier in Norway. Indian J. Meteorol. Geophys., v. 24, n. 3, p. 223-234.

Behrens, H., Bergmann, H., Moser, H., Rauert, W., Stichler, W., Ambach, W., Eisner, H., and Pessl, K., 1971: Study of the discharges of alpine glaciers by means of environmental isotopes and dye tracers. Z. f. Gletscherkunde u Glazialgeologie, v. 7, p. 79-102.

Forsgren, B., 1966: Tritium determinations in the study of palso formation. Geogr. Ann. Ser. A, v. 48a, n. 2, p. 102-110.

Palsa = per glacial mound, Sweden

Izrael, Yu. A., Kolechnikova, V. N., Romanov, V. V., and Soyner, Y. N., 1964: Osoderzhanii tritiya v lednikakh (On the tritium content in glaciers). Akad. Nauk SSSR Doklady, v. 156, n. 1, p. 72-73.

USSR, Parmir Mountains

Krouse, H. R., 1970: Application of isotope techniques to glacier studies (in Glaciers). Ottawa, Secr. Can. Natl. Comm. Internat. Hydrol. Decade, p. 49-50.

Lorius, C., 1964: Isotopes in relation to polar glaciology. Polar Rec., v. 12, n. 77, p. 211-228.

Pb-210, Carbon-14, Sr-90, Deuterium and Oxygen-18

Merlivat, L., Ravoire, J., Vergnaud, J. P., and Lorius, C., 1973: Tritium and deuterium content of the snow in Greenland. Earth Planet. Sci. Lett., v. 19, n. 2, p. 235-240.

Merlivat, L., and Lorius C., 1973: Teneur en deuterium et tritium de la neige du Groenland (The deuterium and tritium contents of snow in Greenland, abstr.). Réun. Annu. Sci. Terre, Programme Résumés, p. 299.

Miller, M. M., Leventhal, J. S., and Libby, W. F., 1965: Tritium in Mt. Everest ice, annual glacier accumulation and climatology at great equatorial altitudes. J. Geophys. Res., v. 70, n. 16, p. 3885-3888.

Moser, H., Rauert, W., Stichler, W., et al.: Messungen des deuterium- und tritiumgehaltes von schnee elsund schmelz wasserproben des Hintereis ferners, Otztaler Alpen (Measurements of deuterium and tritium contents of snow, ice and meltwater from Hintereis Glacier, Otztal Alps). Z. Gletscherk. Glazialgeol., v. 8, n. 1-2, p. 275-281.

Oeschger, H., Renaud, A., and Schumacher, E., 1962: Essai de datation par la tritium des couches de névé du Jungfraufirn et détermination de l'accumulation annuelle (Attempt at dating by tritium of the snow layers of the Jungfraufirn and determination of the annual accumulation). Soc. Vaudoise des Sci. Nat. Bull., v. 68, pt. 1, n. 306, p. 49-56.

Switzerland and Greenland

Picciotto, E., 1961: Quelques résultats scientifiques de l'éxpedition antarctique belge 1957-1958 (Some scientific results of the Belgian Antarctic Expedition, 1957-1958). Ciel et Terre, v. 77, n. 4-6, p. 126-166.

Sör Rondane Mountains

Picciotto, E., 1962: Notes on isotope glaciology. Polar Record, v. 11, n. 71, p. 206-208.

Picciotto, E., 1968: Determination of the rate of snow accumulation at the pole of relative inaccessibility, eastern Antarctica--a comparison of glaciological and isotopic methods. J. Glaciology, v. 7, n. 50, p. 273-287.

Oxygen-18, Pb-210

Ravoire, J., Lorms, C., Robert, J., and Roth, E., 1970: Tritium content in the firn core from Antarctica. J. Geophys. Res., v. 75, n. 12, p. 2331-2335.

Renaud, A., 1969: Observations sur la glace bordière (<u>in</u> Etudes physiques et chimiques sur la glace de l'Indlandsis du Groenland 1959). Medd. Grönland, v. 177, n. 2, p. 107-110.

Shen, S. P., Korff, S. A., and Neuberg, H. A. C., 1963: Tritium content of Antarctic snow. Nature, v. 199, n. 4888, p. 60-61.

Ross Ice Shelf

Smith, J. L., Willen, D. W., and Owens, M. S., 1967: Isotope snow gages for determining hydrologic characteristics of snow packs. Am. Geophys. Union, Geophysical Mon. Ser. n. 11, p. 11-21.
New Zealand

Taylor, C. B., 1964: Tritium content of Antarctic snow. Nature, v. 201, n. 4915, p. 146.

Theodórsson, P., 1971: Rannsóknir á Bárdarbungu 1969 og 1970 (Investigations on Bárdarbungu in 1969 and 1970, English summary). Jökull, v. 20, n. 1970, p. 1-14.
Deuterium, Iceland

Vergnaud, J. P., Botter, R., Lorius, C., et al.: Tritium and deuterium content of snow in Greenland (in Symposium on Hydrogéochemistry and Biogeography, v. 1, Hydrogeochemistry). Washington, D.C., Clarke Co., p. 621-632.

Von Buttlar, H., and Wiik, B., 1965: Enrichment of tritium by thermal diffusion and measurement of dated Antarctic snow samples. Science, v. 149, n. 3690, p. 1371-1373.

Von Buttlar, H., and Wiik, B., 1966: Enrichment of tritium by thermal diffusion and measurement of some well-dated Antarctic snow samples (in Internat. Conf. Radiocarbon and Tritium Dating, 6th Proc., 1965). U.S. Atomic Energy Comm., Rpt. CONF-650652, p. 515-524.

I. OCEANS

Averbach, N. V., 1959: Opredelenie elementov poverkhnostnykh morskikh techenii metodom radioaktivnykh indikatorov (Determination of the elements of surface ocean currents by means of radioactive indicators). Meteorologiia i Gidrologiia, n. 9, p. 41-45.

Bainbridge, A. E., 1963: Tritium in the North Pacific surface water. J. Geophys. Res., v. 68, n. 13, p. 3785-3889.

Bainbridge, A. E., 1963: Tritium in the surface waters of the North Pacific. U.S. Natl. Acad. Sci., Natl. Research Council Publ. 1075, Nuclear Sci. Ser. Rpt. n. 38, p. 129-137.

Bowen, V. T., and Roether, W., 1973: Vertical distributions of strontium 90, cesium 137 and tritium near 45° North in the Atlantic. J. Geophys. Res., v. 78, n. 27, p. 6277-6285.

Dockins, K. O., Bainbridge, A. E., Houtermans, J. C., and Suess, H. E., 1966: Tritium in the mixed layer of the North Pacific Ocean. NTIS, PB-174 711, 14 p.

Dockins, K. O., Bainbridge, A. E., Houtermans, J. C., and Suess, H. E., 1967: Tritium in the mixed layer of the North Pacific Ocean (in Radioactive Dating and Methods of Low-Level Counting--IAEA-ICSW Symposium, Monaco, 1967). Vienna, IAEA, p. 129-141.

Ericson, D. B., 1957: Thick source alpha activity, Fletcher's Ice Island core no. 4 (in Age Measurements and Other Isotopic Studies on Arctic Materials). Columbia Univ., Lamont Geological Observatory, Contract AF-19(604)-1063, 3 p.

Giletti, B. J., 1957: Tritium observations from T-3 Ice Island (in Age Measurements and Other Isotopic Studies on Arctic Materials). Columbia Univ., Lamong Geological Observatory, Contract AF 19 604)-1068, 19 p.

Jenkins, W. J., Clarke, W. B., and Craig, H., 1974: The distribution of ^{3}He in the Northwest Atlantic Ocean (abstr). Eos, Am. Geophys. Union, v. 55, n. 4, p. 313.

Kautsky, H., 1973: Erste ergebnisse der 27 forschungsfahrt von FS "Meteor" (Preliminary results of the 27th voyage of the research vessel Meteor, English summary). Umschau v. 73, n. 3, p. 86-87.

North Atlantic, Bering Sea

Kincaid, G. P., Jr., 1971: Contemporary sources and geochemistry of tritium in the Gulf of Mexico and its distributive province. Texas A&M Univ., Ph.D. dissertation, 243 p., 1972 Diss. Abst. Ind., v. 32, n. 8.

Kincaid, G. P., Jr., and Sackett, W. M., 1971: Tritium levels in the Gulf of Mexico (abstr). Eos, Am. Geophys. Union, Trans., v. 52, n. 4, p. 255.

Martin, J. M., Kulbicki, G., and DeGroot, A. J., 1970: Terrigenous supply of radioactive and trace elements to the ocean (abstr). Int. Symposium Hydrogeochemistry and Biogeochemistry, p. 42.

Merlivat, L., and Menache, M., 1973: Etude de profils de deuterium et de tritium en Meditérranée occidentale (Deuterium and tritium profiles in the western Mediterranean Sea, abstr). Réun. Annu. Sci. Terre, Programme Résumés, p. 300.

Michel, R., and Williams, P. M., 1973: Bomb-produced tritium in the Antarctic Ocean. Earth Planet. Sci. Lett., v. 20, n. 3, p. 381-384.

Molcard, R., Cortecci, G., and Noto, P., 1973: Isotopic analysis of the deep staircase structure in the Tyrrhenian Sea. NTIS, AD-765 815/4, 13 p.

Münnich, K. O., and Roether, W., 1970: Tritium depth profile of the GEOSECS intercalibration cruise station (abstr). Eos, Am. Geophys. Union, Trans., v. 51, n. 4, p. 324.

Ostlund, H. G., Rinkel, M. O., and Rooth, C., 1969: Tritium in the equatorial Atlantic current system. J. Geophys. Res., v. 74, n. 18, p. 4535-4543.

Ostlund, H. G., Dorsey, H., Gorman, and Rooth, C., 1974: GEOSECS North Atlantic radiocarbon and tritium results (in Geochemical Oceans Section Program, GEOSECS, collected papers, 1972-1973. Earth Planet. Sci. Lett., v. 23, n. 1, p. 69-86.

Picciotto, E. E., 1960: Geochemie des éléments radioactifs dans l'ocean et chronologie des sédiments océaniques (Geochemistry of the radioactive elements in the ocean and chronology of the ocean sediments). Ciel et Terre, v. 76, n. 3-4, p. 65-86.

Revelle, R., Folsom, T. R., Goldberg, E. D., and Isaacs, J. D., 1955: Nuclear science and technology (in International Conference on the Peaceful Uses of Atomic Energy, Proceedings). A/CONF 8/P/277, USA, 22 p.

Roether, W., Münnich, K. O., and Ostlund, H. G., 1970: Tritium profile at the North Pacific GEOSECS intercalibration station. J. Geophys. Res., v. 75, n. 36, p. 7672-7675.

Roether, W., and Münnich, K. O., 1972: Tritium profile at the Atlantic 1970 GEOSECS test cruise station. Earth Planet. Sci. Lett., v. 16, n. 1, p. 127-130.

Roether, W., 1974: The tritium carbon-14 profiles at the GEOSECS I(1969) and GOGO I(1971) North Pacific Stations (in Geochemical Ocean Sections Program, GEOSECS, collected papers 1972-1973). Earth Planet. Sci. Lett., v. 23, n. 1, p. 108-115.

Scripps Institution of Oceanography, 1974: Physical and chemical data from the Nemo expedition in the Ross Sea aboard USCGC Northwind (ice breaker) February 1972. NTIS, AD/A-000 799/75L, 29 p.

Sudertund, R., 1971: On the exchange of tritium in the Gulf of Bothnia. Stockholm Univ. Institute Meteorology, Rpt. AC/16, 15 p.

Tamuly, A., 1974: Dispersal of tritium in southern ocean waters (French, Russian summaries). Arctic, v. 27, n. 1, p. 27-40.

Trier, R. M., Broecker, W. S., and Feely, H. W., 1972: Radium-228 profile at the second GEOSECS intercalibration station, 1970, in the North Atlantic. Earth Planet. Sci. Lett., v. 16, n. 1, p. 141-145.

J. SOIL WATER AND UNSATURATED FLOW

Isaacson, R. E., Brownell, L. E., Nelson, R. W., and Roetman, E. L., 1974: Soil moisture transport in arid site vadose zones. Atlantic Richfield Hanford Co., Rpt. ARH-SA-169, 25 p.

Jordan, C. F., Koranda, J. J., Kline, J. R., and Martin, J. R., 1970: Tritium movement in a tropical ecosystem. Bioscience, v. 20, n. 14, p. 807-812.

Jordan, C. F., Kline, J. R., and Sasscer, D. S., 1971: Tritium movement in an old-field ecosystem determined experimentally (in 3rd National Symposium on Radioecology, May 10-12, 1971, Oak Ridge, Tennessee). CONF-710501-P1, p. 199-203.

Kline, J. R., and Jordan, C. F., 1968: Tritium movement in soil of tropical rain movement. Science, v. 160, n. 3827, p. 550-551.

Sasscer, D. S., Jordan, C. F., and Kline, J. R., 1971: Mathematical model of tritiated and stable water movement in an old-field ecosystem (in 3rd National Symposium on Radioecology, May 10-12, 1971, Oak Ridge, Tennessee). CONF-710501, p. 915-923.

Sasscer, D. S., Jordan, C. F., and Kline, J. R., 1973: Dynamic model of water movement in soil under various climatological conditions (in Tritium). Phoenix, Arizona, Messenger Graphics, p. 485-495.

Smith, D. B., Wearn, P. L., Richards, H. J., and Rowe, P. C., 1970: Water movement in the unsaturated zone of high and low permeability strata by measuring natural tritium (in Isotope Hydrology, 1970 Symposium of International Atomic Energy Agency and UNESCO, May 9-13, 1970 Vienna). Vienna, IAEA, STI/PUB/255, Paper n. SM-129/6, p. 73-87.

Smith, J. L., 1972: Forest soils and the associated soil-plant water regime (in Symposium on the Use of Isotopes and Radiation in Research on Soil-Plant Relationships Including Applications in Forestry, December 13-17, 1971, Vienna). Vienna, IAEA, SM-151, p. 399-412.

Stewart, G. L., 1973: The behavior of tritium in the soil (in Tritium). Phoenix, Arizona, Messenger Graphics, p. 462-470.

Wallo, E. W., and Remson, I., 1963: Applications of tracers in studies of soil water movement. Am. Geophys. Union, Trans., v. 44, n. 2, p. 576-577.

Zimmerman, U., Ehhalt, D., and Münnich, K. O., 1967: Soil-water movement and evapotranspiration, changes in the isotopic composition of the water (in Isotopes in Hydrology, Symposium of International Atomic Energy Agency and International Union Geodesy and Geophysics, November 14-18, 1966, Vienna). Vienna, IAEA, STI/PUB/141, paper n. SM-83/38, p. 567-585.

K. GROUND WATER

Areal Studies

Allison, G. B., and Holmes, J. W., 1973: The environment tritium concentration of underground water and its hydrological interpretation. J. Hydrol., v. 19, n. 2, p. 131-143.

Australia (Gambier Plain and Eyre Peninsula)

Barnes, J., and Carswell, D., 1973: Tritium content of underground water in the Wagga area. Geol. Soc. Aust. J., v. 20, pt. 2, p. 175-177.

Australia (New South Wales)

Bedmar, A. P., Sanchez, W., and Castegnat, A. C., et al., 1973: Estudo dos aquiferos das bacias dos Rios Gurgueia e Fidalgo, Piaui, utilizando isotopos ambientais (The study of the aquifers of the basins of Gurgueia River and Fidalgo River, Piaui, Brazil, using ambient isotopes, abstr) (in Resumo das comunicacoes, sessoes techicas). Geofisica, Congr. Bras. Geol., n. 27, Bol. 1, p. 189.

Bennett, R., 1969: Carbon-14 and tritium investigation of ground water in the Tucson Basin. Geol. Soc. Amer. Spec. Paper 121, p. 587.

Bergstrom, J., 1974: Seasonal variations and distribution of dissolved iron in an aquifer. Nordic Hydro, v. 5, n. 1, p. 1-31.

Blavoux, B., et al., 1966: Hydrodynamics and tritium content of waters in Evian Basin, France. Geochem. Internat. v. 2, n. 1, p. 171-173.

Bortolami, G., Fontes, J.-C., and Panichi, C., 1973: Isotopes du milieu et circulations dans les aquifères du sous-sol Venetian (Ambient isotopes and circulation in the aquifers of Venice). Earth Planet. Sci. Lett., v. 19, n. 2, p. 154-167.

Bowen, R., and Williams, P. W., 1972: Tritium analysis of ground water from Grot Lowland of Western Ireland. Experientia, v. 28, n. 5, p. 497-498.

Bredenkamp, D. B., and Vogel, J. C., 1970: Study of a dolomitic aquifer with carbon-14 and tritium (<u>in</u> Isotope Hydrology, 1970 Symposium of International Atomic Energy Agency and UNESCO, March 9-13, 1970). Vienna, IAEA, STI/PUB/255, paper n. SM-129/21, p. 349-372.

South Africa, Transvaal

Brezganov, V. S., Sultankhodzhayez, A. N., and Tyminskiy, V. G., 1967: Primeneniye yestest vennykh radioaktivnykh i stabil'nykh izotopov dlya vyyasneniya usloviy formivaniya mineral nykh vod, na primere Pritashkentskogo artezianskogo basseyna (Use of natural radioactive and stable isotopes to bring out the conditions of formation of mineral waters, exemplified by the Tash Kent Artesian Basin). Uzbek. Geol. Zuhr., n. 5, p. 61-68.

Carlston, C. W., Thatcher, L. L., and Rhodehamel, E. C., 1960: Tritium as a hydrologic tool. The Wharton Tract Study, Internat. Assoc. Sci. Hydrol., Pub. 52, p. 503-512.

Davis, G. H., Gattinger, T. E., Payne, B. R., et al., 1967: Jahreszeitliche schwankungen des tritium gehaltes von grundwässern des Wiener beckens (Seasonal variations in the tritium content of ground waters in the Vienna basin). Austr. Geol. Bundesandt., Verh., n. 1-2, p. 212-232.

Davis, G. H., Payne, B. R., Dincer, T., Florkowski, T., and Gattinger, T., 1967: Seasonal variations in the tritium content of ground waters of the Vienna Basin, Austria (<u>in</u> Isotopes in Hydrology, Symposium of International Atomic Energy Agency and International Union Geodesy and Geophysics, November 14-18, 1966, Vienna). Vienna, IAEA, STI/PUB/141, paper n. SM-83/32, p. 451-473.

Davis, G. H., Meyer, G. L., and Yen, C. K., 1968: Isotope hydrology of the artesian aquifers of the Styrian Basin, Austria (French, German summaries). Steirische Beitr. Hydrogeol., v. 20, p. 51-62.

Davis, G. H., et al., 1970: Geohydrologic interpretations of a volcanic island from environmental isotopes. Water Resources Research, v. 6, n. 1, p. 99-109.

Korea, Cheju Island

Deininger, D. T., 1974: An investigation of ground water in northeastern Florida and southeastern Georgia by analysis of its tritium content (in Southeastern Section, 23rd Annual Meeting). Geol. Soc. Am. Abstr., v. 6, n. 4, p. 349.

Dincer, T., Payne, B. R., Yen, C. K., et al., 1972: Das Tote Gebirge als entwaesserungstypas der karstmassive der nordoestlichen kalkhochalpen, Ergebnisse von isotopenmessungen (The Tote Gebirge as drainage models for the karst drainage of northeastern limestone Alps, isotope measurements). Steirische Beitr. Hydrogeol., v. 24, p. 71-109.

Drost, W., 1970: Ground water measurements at the site of the Sylvenstein Dam in the Bavarian Alps (in Isotope Hydrology, 1970 Symposium of International Atomic Energy Agency and UNESCO, March 9-13, 1970, Vienna). Vienna, IAEA, STI/PUB/255, paper n. SM-129/25, p. 421-437.

Edmunds, W. M., 1973: Trace element variations across an oxidation-reduction barrier in a limestone aquifer (in Symposium on Hydrogeochemistry and Biochemistry, v. 1). Hydrogeology, p. 500-526.

Feder, G. L., 1973: A conceptual model of the hydrologic system supplying the large springs in the Ozarks. Rolla, Missouri, U.S. Geol. Survey, open file rpt., 148 p.

Geyh, M. A., 1972: Basic studies in hydrology and ^{14}C and ^{3}H measurements (in Hydrogeology, Section II). Int. Geol. Congr. Geol. Int. Programme, n. 4, p. 227-234.

Central Europe and Brazil

Giancarlo, B., Fontes, J.-C., and Panichi, C., 1971: Hydrogéologie isotopique de la plaine de Venise; resultats et perspectives (Isotopic hydrogeology of the Venetian plain; preliminary results and perspectives) (in Convegna Internazionale sulle ocque sotterranee). Atti, p. 797-803.

Gunay, G., 1972: Determination of the origin of Ovacik submarine springs by means of natural isotopes (in Congress of Tokyo, Asian Regional Conference, Reunion of Tokyo, Conference Regionale de l'Asie). Internat. Hydrogeol. Mem., v. 9, p. 136-139.

Haskell, E. E., Leventhal, J. S., and Bianchi, W. C., 1966: The use of tritium to measure the movement of ground water toward irrigation wells in western Fresno County. J. Geophys. Res., p. 3849-3859.

Holmes, C. R., 1963: Tritium studies, Socorro Springs (in New Mexico Geological Society Guidebook of the Socorro Region). Socorro, New Mexico, p. 152-154.

Hufen, T. H., Buddmeier, R. W., and Lau, L. S., 1974: Radiocarbon, carbon-13 and tritium in water samples from basaltic aquifers and carbonate aquifers on the island of Oahu, Hawaii. IAEA Symposium, Vienna, February 1974, Paper n. IAEA/SM-132/33, 18 p.

Kaufman, M. E., Rydel, H. S., and Osmund, J. K., 1968: Radiometric studies of the Floridan aquifer, Project completion report (CHAP) B-U-234/U-238 disequilibrium as an aid to hydrologic study of the Floridan aquifer. Washington, D.C., U.S. Office of Water Resources Research, p. B1-B5.

Kohout, F. A., and Stewart, G. L., 1968: Bomb-produced tritium as a tracer of salt water encroachments. Am. Geophys. Union, Trans., v. 49, n. 1, p. 165-166.

Konig, R., and Reitinger, J., 1970: Comparison of hydrological and mathematical methods in analysis of tritium data (in Isotope Hydrology, 1970 Symposium of International Atomic Energy Agency and UNESCO, March 9-13, 1970, Vienna). Vienna, IAEA, STI/PUB/255, Paper n. SM-129/50, p. 787-800.

Leventhal, J. S., Haskall, J. S., and Bianchi, W. C., 1966: The use of tritium to measure the movement of ground water irrigation wells in western Fresno County, California. J. Geophys. Res., v. 71, n. 16, p. 3349-3359.

Mandel, S., Gilboa, Y., and Mercado, A., 1972: Ground water flow in calcareous aquifers in the vicinity of Barcelona, Spain. Internat. Assoc. Hydrol. Sci., Assoc. Internat. Sci. Hydrol. Bull., v. 17, n. 1, p. 77-83.

Margrita, R., Evin, J., and Paloc, H., 1970: Contrition des mesures isotopiques a l'étude de la Fontaine de Vaucluse (Contribution of isotope measurements to the study of Fontaine-de-Vaucluse) (in Isotope Hydrology, 1970 Symposium of International Atomic Energy Agency and UNESCO, March 9-13, 1970, Vienna). Vienna, IAEA p. 333-348.

Matthess, G., Thilo, L., Roether, W., and Münnich, K. O., 1968: Tritium content in the water of low aquifers (preliminary report). Gas-u. Wassfach, v. 109, p. 353-355.

Germany, Rhine Valley

Mazor, E., and Molcho, M., 1972: Geochemical studies on the Feshcha Springs, Dead Sea Basin. J. Hydrol., v. 15, n. 1, p. 37-47.

Mazor, E., Nadler, A., and Molcho, M., 1973: Mineral springs in the Suez Rift Valley; comparison with water in the Jordan Rift Valley and postulation of a marine origin. J. Hydrol., v. 20, n. 4, p. 289-309.

Milde, G., Fröhlich, K., and Klinger, C., 1970: Ergebnisse und praktische folgerungen physikalischer alterbestimmungen au grundwässern (Results and practical consequences of physical age determinations on ground waters) with English and Russian summaries. Zeitschr. Angew. Geologie, v. 16, n. 1, p. 35-40.

Milde, G., Klinger, C., and Fröhlich, K., 1974: Methodische ergebnisse laengerzeitiger tritiumuntersuchungen an grundwaessern aus klueftigporoesen und klueftig-(kavernoesen) grundwasserleitertypen (Methodical results of long-time tritium studies in ground water from jointed-porous and jointed-cavernous types of aquifers) with English and Russian summaries. Zeitschr. Angew. Geologie, v. 20, n. 3, p. 103-107.

Ochiai, T., 1968: Groundwater hydrology in Kanto plain by the use of tritium dating (Japanese and English summaries). Japan Assoc. Groundwater Hydrol. J., n. 14 (v. 43, n. 7), p. 11-23.

Osmund, J. K., and Buie, B. F., 1968: Radiometric studies of the Floridan aquifer. Office of Water Resources Research, Project Completion Rpt. A-005-FLA, 40 p.

Poland, J. F., 1973: New tritium data on movement of ground water in western Fresno County, California (in Fall Annual Meeting 1973, San Francisco, Section of Hydrology, Soil and Ground Water). Eos, Am. Geophys. Union, Trans., v. 54, n. 11, p. 1077-1078.

Purtyman, W. D., 1974: Dispersion and movement of tritium in shallow aquifer in Mortandad Canyon at the Los Alamos Scientific Laboratory. Los Alamos Scientific Laboratory Rpt. LA-5716-MS (NTIS), 10 p.

Reeder, H. O., 1964: Tritium content as an indicator of age and movement of ground water in the Roswell Basin, New Mexico. U.S. Geol. Survey Prof. Paper 501-C, p. C161-C163.

Trilla, A., and Olive, P., 1971: Cronologia de las aguas subterraneas, una aportacion a la hydrogeologia (Chronology of ground water, uses in hydrogeology). Hydrogeologia, n. 1, sec. 3, v. 2, p. 695-702.

Tsurumaki, M., 1972: Water quality problem in eastern Osaka and its hydrological interpretation (Japanese and English summaries). Tokyo J. Geogr., v. 81, n. 5, p. 33-48.

Tsurumaki, M., 1972: Water quality and tritium content of ground water in eastern Osaka, Japan. Univ. Osaka, J. Geosci., v. 15, p. 99-112.

Von Buttlar, H., 1959: Ground water studies in New Mexico using tritium as a tracer, II. J. Geophys. Res., v. 64, n. 8, p. 1031-1038.

Wallick, E. I., 1969: Tritium hydrology of the Tallahassee, Florida area with analysis by liquid scintillation counting without isotopic enrichment. M.S. Thesis, Florida State Univ., 110 p.

Wurzel, P., and Ward, P. R. B., 1969: Ground water studies in the Sabi Valley, Rhodesia, using natural tritium measurements. J. Hydrol., v. 8, n. 1, p. 48-58.

Wurzel, P., 1972: Radioisotopes in underground water investigations in Rhodesia. Geol. Soc. S. Africa., Trans. v. 75, pt. 1, p. 5-10.

Zojer, H., and Zoetl, J., 1974: Die bedeutung von isotopenmessungen im rehmen kombinierter karstwasseruntersuchungen (The significance of isotopic measurements for combined karst water studies). Oesterreichische Wasserwirtschaft, v. 26, n. 3/4, p. 62-70.

Ground Water Recharge

Allison, G. B., and Hughes, M. W., 1972: Comparison of recharge to groundwater under pasture and forest using environmental tritium. J. Hydrol., v. 17, n. 1-2, p. 81-95.

Australia, Gambier Plain

Allison, G. B., Holmes, J. M., and Hughes, M. W., 1973: An investigation of recharge to the northern Adelaine Plains aquifers using environmental tritium. Geol. Soc. Australia, J., v. 19, pt. 4, p. 497-500.

Bredenkamp, D. B., Schutte, J. M., and DuToit, G. J., 1974: Recharge of a domitic aquifer as determined from tritium profiles (in Isotope Techniques in Ground Water Hydrology). Vienna, IAEA, v. 1.

Datta, P. S., Goel, P. S., Rama, et al., 1973: Ground water recharge in western Uttar Pradesh. Indian Acad. Sci., Proc., Sect. A., v. 78, n. 1, p. 1-12.

Dincer, T., Al-Mugrin, A., and Zimmerman, U., 1974: Study of the infiltration and recharge through the sand dunes in arid zones with special reference to the stable isotopes and thermonuclear tritium. J. Hydrol., v. 23, p. 79-109.

Gat, J. R., and Tzur, Y., 1967: Modification of the isotopic composition of rainwater by processes which occur before groundwater recharge (in Isotopes in Hydrology, Symposium of International Atomic Energy Agency and International Union Geodesy and Geophysics, November 14-18, 1966, Vienna). Vienna, IAEA, p. 49-60.

Israel study

Martinec, J., Siegenthaler, U., Oeschger, H., and Tongiorg, E., 1974: New insights into the runoff mechanism by environmental isotopes (in Isotopes Techniques in Ground Water Hydrology). Vienna, IAEA, v. 1.

Switzerland

Martinec, J., 1975: Subsurface flow from snow melt traced by tritium. EOS, Am. Geophys. Union, v. 56, n. 5.

Molina Berbeyer, R., 1963: Uso del tritio natural en la determinación del volumen medio infiltrado de aguas meteoricas en los acuiferos de la subcuenca de Chalco, Mexico (Use of natural tritium for determination of the average volume of meteoric waters in infiltrating into the aquifers of the subbasin of Chalco, Mexico), English abstr. Mexico Univ. Nac. Autónoma Inst. Geofísica Anales, v. 9, p. 103-108.

Comprehensive, 1200 samples, 2 years

Sukhiya, B. S., and Rama, 1973: Evaluation of ground water recharge in semi-arid region of India using environmental tritium. Indian Acad. Sci. Proc., Sect. A. v. 77, n. 6, p. 279-292.

Verhagen; B. Th., Mazor, E., and Sellschop, J. P. F., 1974: Radiocarbon and tritium evidence for direct rain recharge to ground waters in the northern Kalahari. Nature, v. 249, n. 5458, p. 643-644.

Botswana

Vogel, J. C., Thilo, L., and Van Dijken, M., 1974: Determination of ground water recharge with tritium. J. Hydrol. v. 23, n. 1-2, p. 131-140.

South Africa

Age Dating

Afanasenko, V. Ye., Morkovkina, I. K., Romanov, V. V., et al., 1973: O vosraste podzemnykh vod khrebta Tas-khayakhtakh i selennyakhskoy nalozhennoy vpadiny (The age of the ground water of the Tas-khayakhtash Range and the Selennyakh Basin). Moscow, Univ., Vestn., Ser. Geol., v. 28, n. 5, p. 105-109.

Constantinescu, T., 1972: Datarea apelor subterane cu ajutorul izotopilor naturali cu aplicare in cercetarea apelor suberane din zone Bucuresti si cimpia Romania (Dating of ground water by natural isotopes; application in Bucharest and Romanian Plain), English summary. Rom. Inst. Meteorol. Hidrol. Stud. Hidrolgeol., v. 9, p. 1-228.

Gaspar, E., and Oncescu, M., 1972: Radioactive tracers in hydrology. Amsterdam, Elsevier Publ. Co., Developments in Hydrology, I,.342 p.

Gehr, M. A., and Kreysing, K., 1973: Sobre a idade das aguas subterraneas no poligono das secas do nordeste brasileiro (The age of the ground water in the basement of northern Brazil), English summary. Rev. Bras. Geocienc., v. 3, n. 1, p. 53-59.

Geyh, M. A., and Wendt, I., 1965: Results of water sample dating by means of the model by Münnich and Vogel (in Radiocarbon and Tritium Dating). Pullman, Washington. Washington State Univ., p. 597-603.

Hanshaw, B. B., Rubin, M., Stewart, G. L., and Friedman, I., 1968: Verification of radiocarbon dating of ground water by means of tritium analyses (abstr). Am. Geophys. Union, Trans., v. 49, p. 166-167.

Florida study

Kimura, S., Wada, M., Kawasaki, H., et al., 1971: Tritium dating of land subsidence in Niigata (with discussion and French summary). Internat. Assoc. Sci. Hydrol., Publ., v. 1, n. 88, p. 185-192.

Japan

Mangano, F., Marcé, A., and Martin, J. M., et al., 1969: Remarques sur l'utilisation des isotopes radioactifs pour la datation des eaux souterraines (Radioactive dating of ground water). Fr. Bur. Rech. Géol. Minières, Bull. (Sér. 2), Sect. 3, n. 3, p. 39-45.

Mook, W. G., 1972: Application of natural isotopes in ground water hydrology. Geologie en Mijnbouw, Netherlands, v. 51, n. 1, p. 131-136.

Münnich, K. O., and Roether, W., 1963: A comparison of carbon-14 and tritium ages of ground water (in Radioisotopes in Hydrology), French and Russian abstracts. Vienna, IAEA, p. 397-404.

Germany

Münnich, K. O., 1968: Isotopen datierung von grundwasser (Isotopic dating of ground water). Naturwissenschaften, v. 55, n. 4, p. 158-163.

Nir, A., 1964: On the interpretation of tritium age measurements of ground water. J. Geophys. Res., v. 69, n. 12, p. 2589-2595.

Oeschger, H., and Gugelmann, A., 1973: Das geophysikalische verhalten der umweltisotope als basis fuer modellrechnungen in der isotopenhydrologie (The geophysical behavior of the environmental isotopes as a basis for model calculations in isotopic hydrology). Oesterreichische Wasserwirtschaft, v. 26, n. 3/4, p. 43-49.

Oeschger, H., Gugelmann, A., Loosli, H., Schotterer, U., Siegenthaler, and Wiest, W., 1974: Dating of ground water (<u>in</u> Isotope Techniques in Ground Water Hydrology, v. II). Vienna, IAEA.

Switzerland

Osmond, J. K., and Buie, B. F., 1971: Uranium and tritium as natural tracers in the Floridan Aquifer. Florida Water Resources Research Center, Publ. n. 14 (NTIS) 66 p.

Roether, W., 1970: Tritium und kohlenstoff-14 im wasserkreislauf (Tritium and carbon-14 in the hydrologic cycle) (<u>in</u> Sonderheft hydrogeochemie). Deut. Geol. Ges. Z. Sonderheft, p. 182-192.

Schroll, E., 1969: Alterbestimmung an waessern (Age determination of water), summary. Tschermaks Mineral. Petrogr. Mitt., v. 13, n. 3-4, p. 345-346.

Siegenthaler, U., 1972: Bestimmung der verweidauer von grundwasser im boden mit radioaktiven um weltisotopen, carbon-14 and tritium (Dating ground water with radioactive isotopes such as carbon-14 and tritium). Gas Wasser Abwasser, v. 52, n. 9, p. 283-290.

Switzerland

Tamers, M. A., and Scharpenseel, 1970: Sequential sampling of radiocarbon in ground water (<u>in</u> Isotope Hydrology, 1970 Symposium of International Atomic Energy Agency and International Union of Geodesy and Geophysics, March 9-13, 1970). Vienna, IAEA, STI/PUB/255/16, p. 241-257.

Venzuela, Germany

Tolstikhin, I. N., and Kamenskiy, I. L., 1969: On the possibility of ground water age determination using tritium-helium-3 method (Russ.). Geokhimiya, n. 8, p. 1027-1029.

Vogel, J. C., 1970: Carbon-14 dating in ground water (in Isotope Hydrology, 1970 Symposium of International Atomic Energy Agency and UNESCO, March 9-13, 1970, Vienna). Vienna, IAEA, STI/PUB/255, paper n. SM-129/15, p. 225-239.

Reactions with Rocks

Ehhalt, D. H., 1973: On the uptake of tritium by soil water and ground water. Water Resources Res., v. 9, n. 4, p. 1073-1074.

Phillips, R. E., and Brown, D. A., 1968: Self-diffusion of tritiated water in montmorillonite and kaolinite clay. Soil Sci. Amer. Proc., v. 32, n. 3, p. 302-306.

Rabinowitz, D., and Holmes, C. R., 1970: Forced exchange of tritiated water with clay (abstr). EOS, Am. Geophys. Union, Trans., v. 51, n. 4, p. 285.

Rabinowitz, D. D., Holmes, C. R., and Gross, G. W., 1973: Forced exchange of tritiated water with clay (in Tritium). Phoenix, Arizona, Messenger Graphics, p. 471-485.

Stewart, G. L., 1967: Fractionation of tritium and deuterium in soil water (in Isotope Techniques in the Hydrologic Cycle). Am. Geophys. Union, Geophysical Mon. Ser., n. 11, p. 159-168.

Stewart, G. L., 1972: Clay-water interaction, the behavior of H-3 and H-2 in adsorbed water, and the isotope effect. Soil Sci. Soc. Amer. Proc., v. 36, n. 3, p. 421-426.

Tracers

Abdullayev, A. A., Khaitov, B. K., Zakhidov, A. Sh., et al., 1968: Die verwendung von tritium als indikator unterirdischer wässer (The use of tritium as a ground water tracer). Z. Angew. Geol., v. 14, n. 2, p. 86-88.

Alekheyev, F. A., Soifer, V. N., Filonov, V. A., and Finkelstein, Y. B., 1958. Experiment on the use of tritium as an indicator for ground water. USSR, Atom. Energ., v. 4, p. 298-301.

Alekseev, F. A., et al., 1960: Use of the tritium isotope of hydrogen in development of oil fields. Geologiya Nefti - Petroleum Geology (trans.), v. 2, n. 12B, p. 1039-1043.

USSR, Grozny area

Armstrong, F. E., Evans, G. C., and Fletcher, G. E., 1971: Tritiated water as a tracer in the dump leaching of copper. U.S. Bureau of Mines Mineral Resources Report of Investigations RI-7510, 39 p.

Atom, 1972: Ground water tracing with radioactive tracers. London, Atom, n. 192, p. 178-180.

Avramescu, E., Albu, M., and Botezatu, N. P., 1971: Analyse de l'écoulement des eaux souterraines a l'aide des radionucleides pour la détermination des parametres hydrogéologiques (Ground water flow analysis by means of isotopes for the determination of hydrogeologic parameters). Meteorol. Hydrol., n. 2, p. 3-15.

Batsche, H., 1970: Die anwendung radioaktiver isotope in der hydrologie (Application of radioactive isotopes in hydrology), abstr. Deut. Geol. Ges. Z., v. 119, p. 580-581.

Bors, J., Czepkowski, M., and Szopka, J., 1969: Single-well isotopic methods in hydrogeological investigations for draining purposes in the Adámov mine (English trans.). Nukleonika, v. 13, n. 4-5, p. 121-130.

Burdon, D. J., Eriksson, E., Payne, B. R., Papadimitropoulus, T., and Papakis, N., 1963: The use of tritium in tracing karst ground water in Greece (in Radioisotopes in Hydrology, Symposium Proc., Tokyo, March 5-9, 1963). Vienna, IAEA, p. 309-320.

Cotecchia, V., and Pirastru, E., 1966: Concerning the use of radioactive tracers analyses using radioactivity and techniques of isotopic fractionation in hydrogeology. Acqua Agric. lg. Ind., v. 44, p. 7-19.

Drost, W., 1971: Grundwassermessungen mit radioaktiven isotopen (Ground water measurements with radioactive isotopes). Geol. Bavarica, n. 64, p. 167-196.

Drost, W., 1972: Grundwassermessungen im bereich eins Mooreinspiengdammes im degenseemoos bei Penzberg, Oberbayern (Ground water studies at the site of a dam in the Degenseemoos Moor near Penzberg, Upper Bavaria) (in Internationale fachtagung zur untersuchung unterirdischer wasser wege, 2nd, vortraege, diskussionen und beitraege). Ger. Bundesanst. Boden forsch. Geol. Landesaemt. Geol. Jahrb., Reihe, C., n. 2, p. 339-350.

Edwards, J. M., and Holter, L. E., 1954: Radioactive isotopes for water input profiles. Oil Gas Jour., v. 53, n. 30, p. 69-70.

Eriksson, E., and Mosett, F., 1962: Sur l'emploi du tritium comme traceur dans des problèmes d'hydrologie souterraine (Use of tritium as a tracer in subterranean hydrology). Bollettino di Geofisica Teorica ed Applicata, Trieste, v. 4, n. 16, p. 357-360.

Galegar, W. C., and DeGeer, M. W., 1969: Measuring subsurface spring flow with radiotracers. ASCE Proc., J. San. Eng. Div., v. 95, n. SA6, paper 6973, p. 1097-1103.

Texas study

Gasper, E., 1973: Method with radiotracers and experiments in hydrocarst structures. Institut de Fizica Atomica, Rpt. IFA-MR-39, 25 p.

Geyh, M. A., 1974: Erfahrungen mit der 14C- und 3H-methode in der angewandten hydrogeologie (Experiences with 14C and tritium methods in applied hydrogeology). Oesterreichische Wasserwirtschaft, v. 26, n. 3/4, p. 49.

Heemstra, R. J., Watkins, J. W., and Armstrong, F. E., 1961: Laboratory evaluations of nine water tracers. Nucleonics, v. 19, n. 1, p. 92 and 94-96.

Horton, J. H., and Ross, D. I., 1960: Uses of tritium from spent uranium fuel elements as a ground water tracer. Soil Sci., v. 90, p. 267-271.

Iwai, S., and Inoue, Y., 1963: A method for the estimation of ground water stratification (in Radioisotopes in Hydrology, Symposium Proc., Tokyo, March 5-9, 1963). Vienna, IAEA.

Kaufman, W. J., and Orlob, G. T., 1956: An evaluation of ground water tracers. Am. Geophys. Union, Trans., v. 37, p. 297-306.

Kaufman, W. J., and Orlob, G. T., 1956: Measuring ground water movement with radioactive and chemical tracers. J. Amer. Water Works Assoc., v. 48, p. 559-572.

Kaufman, W. J., 1960: Tritium as a gound water tracer. ASCE, Proc., J. Sanitary Eng., Div., p. 47-57.

Kaufman, W. J., and Todd, D. K., 1962: Application of tritium tracers to canal seepage measurements (in Symposium, Tritium in the Physical and Biological Sciences). Vienna, IAEA, p. 83-94.

Keeley, J. W., and Scalf, M. R., 1969: Aquifer storage determination by radiotracer techniques. Ground Water, v. 7, n. 1, p. 17-22.

Knight, A. H., Boggie, R., and Shepherd, H., 1972: The effect of ground water level on water movement in peat, a study using tritiated water. J. Appl. Ecol., v. 9, n. 2, p. 633-641.

Knutsson, G., Ljunggren, K., and Forsberg, H. G., 1963: Field and laboratory tests of chromium-51-EDTA and tritium water as a double tracer for ground water flow (in Radioisotopes in Hydrology, Symposium Proceedings, Tokyo, March 5-9, 1963). Vienna, IAEA, p. 347-363.

Knutsson, G., 1964: Cr-51-EDTA and other tracers for measuring ground water flow. Grundförbättring, v. 17, p. 145-156.

Kriz, G. J., Lewis, D. C., and Burgy, R. H., 1965: Ground water tracing by tritiated water injection. Amer. Sci. Agric. Engrs., Prep. Pap. n. 64, p. 821.

Krotowicz, J., 1967: Izotopowe wskaźniki przeplywu wód podziemnych (Tracing of water flow by means of radioactive isotopes), English, Russian summaries. Przegl. Geol., v. 15, n. 1, p. 30-35.

Kuyper, E., Hofstra, A., and Nauta, H., 1962: Application of tritiated water as a tracer for quantitative determination of water flow distribution in an oil field.

(in Radioisotopes in the Physical Sciences and Industry Vienna, IAEA, p. 511-517.

Lallemand, A., and Grison, G., 1970: Contribution à la sélection de traceurs radioactifs pour l'hydrogéologie (The selection of radioactive tracers for hydrogeology) with discussion and English summary (in Isotope Hydrology, 1970 Proceedings). Vienna, IAEA, p. 823 833.

Lenda, A., and Zuber, A., 1970: Tracer dispersion in ground water experiments (in Isotope Hydrology, 1970 Proceedings). Vienna, IAEA, p. 619-641.

Lewis, D. C., and Burgy, R. H., 1964: The relationship between oak tree roots and ground water in fractured rock as determined by tritium tracing. J. Geophys. Res. v. 69, n. 12, p. 2579-2588.

California, Placer County

Mairhofer, J., and Radl, F., 1971: Ein beispiel zur feststellung von grundwasserstroemungen durch die einlochmethode und tritiumanalysen (Determination of ground water flow by the dilution method and by tritium analysis) with English summary. Steirsche Beitr. Hydrogeol., v. 23 p. 117-126.

Marine, I. W., 1967: The use of tracer test to verify an estimate of the ground water velocity in fractured crysttaline rock at the Savannah River plant near Aiken, S.C. (in Isotope Techniques in the Hydrologic Cycle, Symposium Univ. Illinois,1965). Am. Geophys. Union, Geophys. Mon. Ser., n. 11, p. 171-179.

Marine, I. W., Webster, D. S., and Proctor, J. F., 1969: Two-well tracer test in fractured crystalline rock (abstr). Geol. Soc. Amer. Abstr., pt. 4 (Southeast Sect), p. 48-49.

Merritt, W. F., 1967: Techniques of ground water tracing using radionuclides (in Isotope Techniques in the Hydrologic Cycle, Symposium, Univ. Illinois, 1965). Am. Geophys. Union., Geophys. Mon. Ser., n. 11, p. 169-170.

Mokady, R. S., and Zaslavsky, D., 1967: Radioactive tracers in diffusion tests. Soil Sci. Soc. Amer., Proc. v. 31, n. 5, p. 604-608.

Morris, D. A., 1967: Use of chemical and radioactive tracers at the national reactor testing station, Idaho (in Isotope Techniques in the Hydrologic Cycle). Am. Geophys. Union, Geophys. Mon. Ser., n. 11, p. 130-142.

Moser, H., 1974: Die verwendung kuenstlicher radioaktiver markierungen zur loesung hydrologischer probleme, teil I. Methodik (The use of artificial radioactive tracers for the solution of hydrological problems, part I, methods). Oesterreichische Wasserwirtschaft, v. 26, n. 3/4, p. 75-80.

Nielson, D. R., and Biggar, J. W., 1967: Radioisotopes and labeled salts in soil water movement (in Panel on the Use of Isotope and Radiation Techniques in Soil Moisture and Irrigation Studies, March 1966, Vienna). Vienna, IAEA, p. 61-76.

Reeder, H. O., 1963: Tritium used as a ground water tracer between Lake McMillan and Major Johnson springs, Eddy County, New Mexico. U.S. Atomic Energy Comm., TEI-839, 139 p.

Smith, D. B., 1961: Ground water tracing. Atom, n. 57, p. 26-27.

Thatcher, L. L., 1961: Evaluation of hydrologic tracers. U.S. Geol. Survey Prof. Pap. 424-D, D.396-D.397.

Theis, C. J., 1963: Hydrologic phenomena affecting the use of tracers in timing ground water flow (in Radioisotopes in Hydrology, Symposium Proceedings, Tokyo, March 5-9, 1963). Vienna, IAEA, p. 193-206.

Trilla Arrufat, J., 1972: Estudio hidrogeologica de la cuenca del Francoli; cronologia de las aguas subterraneas (Hydrology of the Francoli River Basin; chronology of ground water), English summary. Acta Geol. Hisp., v. 7, n. 5, p. 138-142.

L. THERMAL WATER

Ariizumi, A., and Kondo, O., 1963: Investigation of the movement of infiltrating acidic hot spring water in the ground by means of radioisotopes (in Radioisotopes in Hydrology, Symposium Proceedings, Tokyo, March 5-9, 1963). Vienna, IAEA, p. 365-381.

Banwell, C. J., 1967: Oxygen and hydrogen isotopes in New Zealand thermal areas (in Nuclear Geology on Geothermal Areas, Spoleto, 1963). Pisa, Italy, Laboratorio Geologia Nucleare, p. 95-138.

Begemann, F., 1967: The tritium content of hot springs in some geothermal areas (in Nuclear Geology on Geothermal Areas, Spoleto, 1963). Pisa, Italy, Laboratorio Geologia Nucleare, p. 55-70.

California, Nevada, Wyoming

Cheminée, J., Létolle, R., and Olive, P., 1969: Premières données isotopiques sur des fumeralles de volcans italiens (First isotopic data on the fumaroles of Italian volcanoes). Bull. Volcanol., v. 32, n. 3, p. 469-475.

Cherdyntsev, V. V., 1973: Yadernaya vulkanologiya (Nuclear volcanology). Akad. Nauk SSSR, Geol. Inst., 208 p.

South Pacific

Devirts, A. L., Komenskiy, I. L., and Tolstikhin, I. N., 1971: Izotopy geliya i tritiy v vulkanicheskikh istochnikakh (Isotopes of helium and tritium in volcanic springs). Akad. Nauk SSSR Doklady, v. 197, n. 2, p. 450-452.

USSR, Kurile Islands and Kanchahka

Gunter, B. D., 1968: Geochemical and isotopic studies of hydrothermal gases and waters. Univ. Arkansas, Ph.D. Dissertation.

California, Wyoming

Jacobshagen, V., and Münnich, K. O., 1964: Carbon-14 age determination and other isotopic investigations on hot salt water springs of the Ruhr carboniferous. Neues Jahrb. Geol., Palaont., Monatsh., n. 9, p. 566-568.

Matsubaya, O., Sakai, H., Dusachi, I., et al., 1973: Hydrogen and oxygen isotopic ratios and major element chemistry of Japanese thermal water systems. Geochim. J. (Geochem. Soc. Japan), v. 7, n. 3, p. 123-151.

Rajner, V., Schroll, E., and Stepan, E., 1967: Tritiummessungen von heissen wassern am strand der insel Volcano, Liparische Inseln (Tritium measurements of the hot waters on the shore of Vulcano Island, Lipari Island). Akad. Wiss. Math. Naturwiss. Kl. Anz., v. 104, p. 58-60.

Italy

Tongiorgi, E. (ed), 1967: Nuclear geology on geothermal areas - Spoleto 1963. Pisa, Italy, Laboratorio Geologia Nucleare, 284 p.

A collection of classic papers

White, D. E., and Craig, H., 1959: Isotope geology of the Steamboat Springs area, Nevada (abstr). Geol. Soc. Amer. Bull., v. 70, n. 12, pt. 2, p. 1696.

Wilson, S. H., 1967: Tritium determinations on bore waters in the light of chloride-enthalpy relations (in Nuclear Geology on Geothermal Areas, Spoleto 1963). Pisa, Italy, Laboratorio Geologia Nucleare, p. 173-184.

Wilson, S. H., 1966: Origin of tritium in hydrothermal solutions. Nature, v. 211, n. 5046, p. 272-273.

New Zealand

12. BIOLOGY AND ECOLOGY

Anonymous, 1973: Kinetics of tritium of biological systems (in Tritium). Phoenix, Arizona, Messenger Graphics, p. 289-381.

Anspaugh, L. R., Koranda, J. J., Robinson, W. L., and Martin, J. R., 1971: Dose to man via food chain transfer resulting from exposure to tritiated water vapor. (Tritium Symposium, Las Vegas, Nevada), Univ. California Lawrence Radiation Lab. Rpt. UCRL-73195 (NTIS), 31 p.

Ballereau, P., 1970: Oxidation of tritium to tritiated water and health physics. Commissariat à l'Energie Atomique (France), Rpt. CEA-B,b-179, 65 p.

Barton, C. J., Moore, R. E., and Rohwer, P. S., 1974: Philosophy and methodology of assessment of radiation doses to populations from combustion products of Plowshare natural gas (in 8th Midyear Topical Symposium of the Health Physics Society, October 21-24, 1974, Knoxville, Tennessee). CONF-741018, p. 369-374.

Battelle-Pacific Northwest Laboratory, 1973: Fresh water ecology. Rpt. BNWL-1750, v. 1, pt. 2, sec. 6, 26 p.

Bennett, B. G., 1972: The radiation dose due to acute intake of tritium by man. U.S. Atomic Energy Comm., Health and Safety Lab., Rpt. HASL-253 (NTIS), 17 p.

Blaylock, B. G., 1971: Chromosome aberrations in *chironomus riparius* developing in different concentrations of tritiated water (in 3rd National Symposium on Radioecology, May 10-12, 1971, Oak Ridge, Tenn). CONF-710501, p. 1169-1173.

Bogen, D. C., Henkel, C. A., White, C. G. C., and Welford, G. H., 1973: Tritium intake in New York (in Tritium). Phoenix, Arizona, Messenger Graphics, p. 630-646.

Bond, V. P., 1970: Evaluation of Potential Hazards from tritiated water (in IAEA-SM-146-13, Proc.). New York, New York, Unipub, Inc., p. 287-300.

Brock, T. D., 1967: Mode of filamentous growth of *leucothrix mucor* in pure culture and in nature as studied by tritiated autoradiography. J. Bacteriology, v. 93, n. 3, p. 985-990.

United States--New York and Rhode Island; Iceland

Brown, R. M., and Lloyd, E. S., 1975: Tritium in vegetation of the Perch Lake disposal area (in Progress Report of Biological and Health Physics Division, October 1-December 31, 1974). Physics Div., Chalk River Nuclear Labs, PR-B-104, item 3.17.

Canada

Bush, W. R., 1972: Assessing and controlling the hazard from tritiated water. Atomic Energy of Canada, Ltd., Rpt. AECL-4150 (NTIS), 106 p.

Cahill, D. F., 1970: Some effects of mammalian developments of continuous irradiation *in utero* with tritiated water. NTIS, UR-49-1254, 86 p.

Cleaver, J. E., Thomas, G. H., and Burki, H. J., 1972: Biological damage from intranuclear tritium; DNA strand breaks and their repair. Science, v. 177, n. 4053, p. 996-998.

Clegg, B., Koranda, J., and Hadley, G., 1971: A system for correlating tritium oxide transport in vegetation with micrometeorological variables. Univ. California Lawrence Radiation Lab., Rpt. UCRL-73373, 6 p.

Eckert, J. A., and Evans, R. B., 1973: Tritium burdens in two Arctic villages. Radiation Data Rpts., v. 14, n. 5, p. 273-275.

Elwood, J. W., 1971: Tritium behavior in fish from a chronically contaminated lake. NTIS, CONF-710501-45, 13 p.

Erikson, R. C., 1971: Effects of chronic irradiation by tritiated water on *Poecilia reticulata*, the guppy. Washington Univ., Lab. Radiation Ecology, Rpt. CONF-710501-38 (NTIS), 27 p.

Euratom Study Group, 1970: Work of the Euratom study group on radiation and human cytogenetics of the University of Pavia. Final report, 1 April 1963-31 March 1968 (in Italian). NTIS, EUR-4526, 20 p.

Fabre, J. L., Henry, P., Mazaury, E., Mercier, J., and Mion, C., 1970: Use of extra-renal purification methods in the treatment of tritiated water contamination (French and English summaries). Commissariat à l'Energie Atomique, Centre de Production de Plutonium, Rpt. CEA-R-3974, 87 p.

Harrison, F. L., and Koranda, J. J., 1971: Tritiation of aquatic animals in an experimental freshwater pool. Univ. California, Lawrence Radiation Lab., Rpt. URCL-72930 (NTIS), 35 p.

Ichikawa, R., and Suyama, I., 1974: Effects of tritiated water on the embryonic development of two marine teleosts. Bull. Jap. Soc. Sci. Fisheries, v. 40, n. 8, p. 819-824.

Kanazawa, T., Kanazawa, K., and Bassham, J. A., 1972: Tritium incorporation in the metabolism of *Chlorella pyrenoidosa*. Env. Sci. Technol., v. 6, n. 7, p. 638-642.

Kelly, M. J., and Booth, R. S., 1971: Transfer of tritium to man from an initial wet deposition. Oak Ridge National Laboratory, Rpt. ORNL-TM-3134 (NTIS), 22 p.

Kirchmann, R., Koch, G., Adam, V., Van Den, and Hoek, J., 1973: Studies on the food chain contamination by tritium (in Tritium). Phoenix, Arizona, Messenger Graphics, p. 341-348.

Kline, J. R., Colon, J. A., Brar, S. S., Stewart, M. L., and Jordan, C. F., 1971: Radiological physics division annual report, environmental research January-December 1971. Argonne National Laboratory, Rpt. ANL-7860 (Pt. 3) (NTIS), 314 p.

Kline, J. R., Jordan, C. F., and Rose, R. C., 1971: Transpiration measurement in pines using tritiated water as a tracer (in 3rd National Symposium on Radioecology, May 10-12, 1971, Oak Ridge, Tenn.). CONF-710510-P1 (NTIS) p. 117-133.

Kline, J. R., and Stewart, M. L., 1972: Tritium uptake and loss in grass vegetation when exposed to an atmospheric source of tritiated water. Argonne National Lab., Rpt. ANL-7960, Pt. 3 (NTIS), p. 117-133.

Kline, J. R., Stewart, M. L., Jordan, C. F., and Kovac, P., 1972: Use of tritiated water for determination of plant transpiration and biomass under field conditions (in Symposium on the Use of Isotopes and Radiation in Research on Soil-Plant Relationships Including Applications in Forestry, December 13-17, 1971, Vienna). Vienna, IAEA, SM-151, p. 419-437.

Koenig, I. A., and Winter, M., 1971: Über die tritiumkontamination der unwelt (Tritium contamination of the environment) (in Commission of the European Communities International Symposium on Radioecology Applied to the Protection of Man and his Environment, September 1971, Rome), 14 p.

Koranda, J. J., and Martin, J. R., 1971: Movement of tritium in ecological systems. Univ. California, Lawrence Radiation Lab., Rpt. URCL-73178 (NTIS), 56 p.

Martin, J. R., Jordan, C. F., Koranda, J. J., and Kline, J. R., 1970: Radioecological studies of tritium movement in a tropical rain forest (in Symposium on Engineering with Nuclear Explosives, January 14-16, 1970, Las Vegas, Nevada). U.S. Atomic Energy Comm., CONF-700101, v. 1 (NTIS).

Martin, J. R., and Koranda, J. J., 1971: Importance of tritium in the civil defense context. Univ. California, Lawrence Radiation Lab., Rpt. URCL-73085 (NTIS), 15 p.

Moghissi, A. A., and Lieberman, R., 1970: Tritium body burden of children, 1967-1968. NTIS, PB-217 641, 5 p.

Moghissi, A. A., Patzer, R. G., and Carter, M. W., 1973: Biokinetics of environmental tritium (in Tritium). Phoenix, Arizona, Messenger Graphics, p. 314-322.

Moore, R., 1962: A comparison of HTO in plasma and expired water vapor. Health Physics, v. 7, n. 3/4, p. p. 161-169.

Moore, R. E., and Barton, C. J., 1973: Dose estimations for the hypothetical use of nuclearly stimulated natural gas in the Cherokee steam electric station, Denver, Colorado. Oak Ridge National Lab., Rpt. ORNL-TM-4026 (NTIS), 39 p.

Morgan, T. J., Landolt, R. R., and Hamelick, J., 1973: Behavior of tritium in fish following chronic exposure (in Tritium). Phoenix, Arizona, Messenger Graphics, p. 378-381.

Moskalev, Yu. I., Shtukkenberg, Yu. M., and Zhuravlev, V. F., 1969: Recommendations concerning the maximum permissible tritium oxide concentrations in the atmosphere of industrial premises. Oak Ridge National Lab., Rpt. ORNL-TR-2403 (NTIS), 7 p.

Myers, D. S., Tinney, J. F., and Gudiksen, P. H., 1971: Health physics aspects of tritium release. Univ. California, Lawrence Radiation Lab., Rpt. UCRL-73310 (NTIS), 18 p.

Osborne, R. V., 1972: Permissible levels of tritium in man and the environment. Radiat. Res., v. 50, n. 1, p. 197-211.

Pennsylvania State University, 1974: Lethal and mutagenic effects in microorganisms grown in exp 3 H sub 2 O, final report. NTIS, COO-3423-4, 3 p.

Planet, J., Uzzan, G., and LeGrand, J., 1970: Radiation exposure to man from disposal of tritiated water vapor to the atmosphere. Commissariat à l'Energie Atomique, Centre d'Etudes Nucléaires (NTIS), 25 p.

Potter, G. D., Vattuone, G. M., and McIntyre, D. R., 1972: Metabolism of tritiated water in the dairy cow. Health Physics, v. 22, n. 4, p. 405-409.

Rohwer, P. S., Kelley, M. J., and Booth, R. S., 1973: Guidance for limiting environmental releases of tritium (in Tritium). Phoenix, Arizona, Messenger Graphics, p. 422-430.

Rosenthal, G. M., Jr., and Stewart, M. L., 1974: Tritium incorporation in algae and transfer in simple aquatic food chains. U.S. Atomic Energy Comm., CONF-710501-24 (NTIS), 16 p.

Seymour, A., 1970: Columbia River studies 1969-1970, effects of temperature and ionizing radiation on the larvae of the Pacific oyster. Washington Univ., Laboratory of Radiation Ecology, RLO-2225-T-1-2 (NTIS).

Skauen, D. M., 1964: The effects of tritium oxide on aquatic organisms, final report June 1, 1962 to July 31, 1964. U.S. Atomic Energy Comm., NYO-3039-1, 18 p.

Smith, D. D., and Giles, K. R., 1974: Animal investigation program 1970 annual report. National Environmental Research Center, Rpt. NERC-LV-539-16 (NTIS), 24 p.

Stewart, G. L., Wyerman, T. A., Sherman, M., and Schneider, R., 1972: Tritium in pine trees from selected locations in the United States, including areas near nuclear facilities. U.S. Geol. Survey, Prof. Paper 800-B, p. B265-B271.

Stewart, M. L., Rosenthal, G. M., and Kline, J. R., 1971: Tritium discrimination and concentration in fresh water microcosms. Argonne National Laboratory (TNIS), 29 p.

Strand, J. A., Templeton, W. L., Tangen, E. G., and Olson, P. A., 1971: Fixation and long-term accumulation of tritium in an experimental aquatic environment and effects of short range particle irradiation on embryogenesis of marine teleost fish. Battelle Pacific Northwest Lab., Rpt. BNWL-1550, v. 1, pt. 2, p. 2.36-2.40.

Strand, J. A., Fujihara, M. P., Templeton, W. L., and Tangen, E. G., 1972: Suppression of *Chondrococcus columnaris* immune response in Rainbow trout sublethally exposed to tritiated water during embryogenesis (in Symposium on the Interaction of Radioactive Contaminants with the Constituents of the Marine Environment, July 10-14, 1972, Seattle, Washington). New York, NY, Unipub Inc., Paper IAEA/SM-158/33, 13 p.

Thompson, R. C., 1971: A review of laboratory animal experiments related to the radiobiology of tritium. Battelle, Pacific Northwest Lab. Rpt. BNWL-SA-3739, 19 p.

Tucker, J. S., and Harrison, F. L., 1974: The incorporation of tritium in the body water and organic matter of selected marine invertebrates. Comparative Biochemistry Physiology, v. 49A, p. 387-397.

Van Den Hoek, J., and Kirchmann, R., 1971: Tritium secretion into cow's milk after administration of organically bound tritium and of tritiated water. Commission of the European Communities International Symposium, Radioecology Applied to the Protection of Man and his Environment, September 1971, Rome, 12 p.

Walden, S. J., 1971: Effects of tritiated water on the embryonic development of the three-spine stickleback, *Gasterosteus aculeatus linneaus*. Washington Univ., Laboratory of Radiation Ecology, Rpt. CONF-710301-37 (NTIS), 13 p.

Wallick, E. I., 1969: Liquid scintillation detection of low-level tritium in ground water of the Tallahassee, Florida area (abstr). Geol. Soc. Amer., Abstr., Pt. 4, (Southeast Sect.), p. 84-85.

Wharton, G. W., 1969: Determination of minimum concentrations of environmental water capable of supporting life, semiannual report, 1 November 1968 - 30 April 1969. Ohio State Univ., Research Foundation, Acarology Lab. (NTIS), 93 p.

Woodward, H. Q., 1970: The biological effects of tritium. Sloan-Kettering Inst. Cancer Research (NTIS), 49 p.

Young, S. E., Dodd, J. D., and Ibert, E. R., 1970: Tritium collection and extraction techniques for plant-water relationship studies. Ecology, v. 51, n. 3, p. 535-537.

Zhuravlev, V. F., 1971: Comparative toxicity of tritium oxide for different animals. Hygiene Sanitation, v. 36, n. 6, p. 380-394.

13. NUCLEAR FACILITIES

A. ENVIRONMENTAL STATUS AND MONITORING OF FACILITIES

General

U.S. Atomic Energy Commission, 1973: Environmental monitoring at major U.S. Atomic Energy Commission contractor sites, calendar year 1972. NTIS, WASH-1259, 1223 p.

Argonne National Laboratory

Sedlet, J., Golchert, N. W., and Duffy, T. L., 1974: Environmental monitoring at Argonne National Laboratory, annual report, 1973. ANL-8078 (NTIS), 88 p.

Steindler, M. J., Levitz, N. M., Trevorrow, L. E., Gering, T. J., and Kullen, B. J., 1973: Chemical engineering division, waste management programs, quarterly report, July-September 1973. ANL-8061 (NTIS), 28 p.

Brookhaven

Hull, A. P., and Ash, J. A., 1974: 1973 environment monitoring report. BNL-18625 (NTIS), 73 p.

Chalk River, Canada

Atomic Energy of Canada, Ltd., 1973: Biology and health physics division, progress report, October 1-December 31, 1972. AECL-4451 (NTIS), 107 p.

Atomic Energy of Canada, Ltd., 1974: Biology and health physics division, progress report, October 1-December 31, 1973. AECL-4765 (NTIS), 99 p.

Mawson, C. A., and Cowper, G., 1972: Chalk River nuclear laboratories progress report, April 1-June 30, 1972, biology and health physics division, environmental research branch and health physics branch. AECL-4272 (NTIS), p. 33-65.

Grenoble, France

Guizerix, J., Margrita, R., Launay, M., and Ruby, P., 1966: Tritium and hydrogeology investigations and measurements carried out at the Grenoble Nuclear Research Center (in Isotopes in Hydrology, Symposium of the International Atomic Energy Agency and International Union Geodesy and Geophysics, November 14-18, 1966, Vienna). Vienna, IAEA, STI/PUB/141, paper n. SM-83/31, 18 p.

Hanford Reservation

Battelle Pacific Northwest Laboratories, 1968: Environmental status of the Hanford Reservation for January-June 1968. BNWL-CC-1850 (NTIS), 68 p.

Bramson, P. E., 1972: Environmental status of the Hanford Reservation for 1970. BNWL-C-96 (NTIS), 102 p.

Bramson, P. E., Corley, J. P., and Nees, W. L., 1973: Environmental status of the Hanford Reservation for CY-1972. BNWL-B-278 (NTIS), 70 p.

Brown, D. J., 1962: Chemical effluent technology waste disposal investigations, July-December, 1961. HW-72645-RD (NTIS), 17 p.

Brown, D. J., 1965: Chemical effluent technology waste disposal investigations, January-June, 1965. BNWL-CC-285 (NTIS), 17 p.

Brown, D. J., and Haney, W. A., 1974: Chemical effluents technology waste disposal investigation, July-December, 1963; the movement of contaminated ground water from the 200 areas to the Columbia River. HW-80909 (NTIS), 17 p.

Denham, D. H., 1969: Radiological status of the ground water beneath the Hanford project, January-June, 1969. BNWL-1233 (NTIS), 20 p.

Denham, D. H., 1970: Radiological status of the ground water beneath the Hanford project, July-December, 1969. BNWL-1392 (NTIS), 20 p.

Eliason, J. R., 1966: Earth sciences waste disposal investigations, July-December, 1965. BNWL-CC-574 (NTIS), 20 p.

Kipp, K. L., Jr., 1973: Radiological status of the ground water beneath the Hanford project, July-December, 1972. BNWL-1752 (NTIS), 29 p.

Kipp, K. L., Jr., 1973: Radiological status of the ground water beneath the Hanford project, January-June, 1972. BNWL-1737 (NTIS), 31 p.

Nees, W. L., and Corley, J. P., 1974: Environmental surveillance at Hanford for CY-1973. BNWL-1811 (NTIS), 85 p.

Routson, R. C., 1973: Review of studies on soil-waste relationships on the Hanford Reservation from 1944-1967. BNWL-1464 (NTIS), 63 p.

Wilson, C. B., and Essig, T. H., 1970: Environmental status of the Hanford Reservation for July-December, 1969. BNWL-CC-2583 (NTIS), 80 p.

Wilson, C. B., and Essig, T. H., 1970: Environmental status of the Hanford Reservation for January-June, 1969. BNWL-CC-2478 (NTIS), 94 p.

Woolridge, C. B., 1969: Environmental status of the Hanford Reservation for July-December, 1968. BNWL-CC-2026 (NTIS), 84 p.

Karlsruhe, Germany

Schieferdeck, H., 1973: Ergebnisse der tritium inkorporationsuberwachung im kernforschungzentrum karlsruhe in den jahren, 1971-1972. KFK-EXT-23/73-1 (NTIS), 12 p.

Lawrence Livermore

Gudiksen, P. H., Lindeken, C. L., Gatrousis, C., and Aspaugh, L. R., 1972: Environmental levels of radioactivity in the vicinity of the Lawrence Livermore Laboratory, January through December, 1971. UCRL-51242 (NTIS), 47 p.

Silver, W. J., Lindeken, C. L., Meadow, J. C., Hutchin, W. H., and McIntyre, D. R., 1974: Environmental level of radioactivity in the vicinity of the Lawrence Livermore Laboratory, 1973 annual report. UCRL-51547 (NTIS), 49 p.

Tinney, J. F., Myers, D. S., Gudiksen, P. H., and Yoder, R. E., 1971: Environmental sampling program following an accidental tritium release. UCRL-51029 (NTIS), 18 p.

Los Alamos Scientific Laboratory

Hakonson, T. E., Johnson, L. J., and Purtyman, W. D., 1973: Ecodistribution of plutonium in waste disposal areas at Los Alamos. LA-UR-73-1001 (NTIS), 23 p.

Purtyman, W. D., 1973: Regional survey of tritium in surface and ground water in the Los Alamos area, New Mexico, August 1966 to May 1969. LA-5234-MS (NTIS), 9 p.

Schiager, K. J., and Apt, K. E., 1974: Environment surveillance at Los Alamos during 1973. LA-5586 (NTIS), 63 p.

Mound Laboratory

Carfagno, D. G., and Westendorf, W. H., 1972: Environmental monitoring report, January-June, 1972. MLM-1973 (NTIS), 18 p.

Carfagno, D. G., and Westendorf, W. H., 1973: Annual environmental monitoring report, calendar year 1972. MLM-2028 (NTIS), 32 p.

Carfagno, D. G., and Westendorf, W. H., 1973: Interim environmental monitoring report, January-June, 1973. MLM-2085 (NTIS), 21 p.

Carfagno, D. G., and Westendorf, W. H., 1974: Annual environmental monitoring report, calendar year 1973. MLM-2142 (NTIS), 37 p.

Mound Laboratory, 1971: Environmental monitoring report, January-June, 1970. MLM-1784 (NTIS), 36 p.

Mound Laboratory, 1972: Environmental monitoring report, January-June, 1971. MLM-1880 (NTIS), 41 p.

Oak Ridge National Laboratory

Compere, E. L., Freid, S. H., and Nestor, C. W., 1974: Distribution and release of tritium in high-temperature gas-cooled reactors as a function of design, operational and material parameters. ORNL-TM-4303 (NTIS), 97 p.

Oak Ridge National Laboratory, 1974: Applied health physics and safety annual report for 1973. ORNL-4974 (NTIS), 56 p.

Union Carbide Corporation, 1974: Environmental monitoring report--United States Atomic Energy Commission, Oak Ridge facilities, calendar year 1973. UCC-ND-280 (NTIS), 46 p.

Pantex Plant

Alexander, R. E., 1974: Environmental monitoring report for Pantex Plant covering 1973. MHSMP-74-12 (NTIS), 44 p.

Pinellas Plant

General Electric Company, 1972: Environmental monitoring report, 1972, Pinellas Plant, St. Petersburg, Florida (in Environmental Monitoring at Major U.S. Atomic Energy Commission Contractor Sites, Calendar Year 1972). Report WASH-1259, p. 177-196.

General Electric Company, 1974: Environmental monitoring report, 1973. TID-26578 (NTIS), 25 p.

Rocky Flats Plant

Werkema, G. J., and Thompson, M. A., 1974: Annual environmental monitoring report, Rocky Flats Plant, January through December 1973. RFP-ENV-73 (NTIS), 37 p.

Savannah River Plant

Ashley, C., and Zeigler, C. C., 1973: Environmental monitoring at the Savannah River Plant, annual report, 1972. DPSPU-73-302 (NTIS).

Du Pont de Nemours (E. I.) and Co., 1959: Health physics regional monitoring semiannual report, January-June, 1959. DPSPU-59-11-30 (NTIS), 41 p.

Du Pont de Nemours (E. I.) and Co., 1960: Health physics regional monitoring semiannual report, July-December, 1959. DPSPU-60-11-9 (NTIS), 44 p.

Du Pont de Nemours (E. I.) and Co., 1960: Health physics regional monitoring semiannual report, January-June, 1960. DPSP-60-25-26(Del), (NTIS), 56 p.

Du Pont de Nemours (E. I.) and Co., 1961: Health physics regional monitoring semiannual report, July-December, 1960. DPSP-61-25-4(Del), (NTIS), 56 p.

Du Pont de Nemours (E. I.) and Co., 1962: Health physics regional monitoring semiannual report, January-June, 1961. DPSP-62-25-2(Del), (NTIS), 68 p.

Du Pont de Nemours (E. I.) and Co., 1962: Health physics regional monitoring semiannual report, July-December, 1961. DPSP-62-25-9(Del), (NTIS), 67 p.

Du Pont de Nemours (E. I.) and Co., 1963: Health physics environmental monitoring semiannual report, January-June, 1962. DPSP-63-25-3(Del), (NTIS), 50 p.

Du Pont de Nemours (E. I.) and Co., 1964: Health physics environmental monitoring annual report, 1963. DPSPU-64-11-12 (NTIS), 39 p.

Du Pont de Nemours (E. I.) and Co., 1966: Tritium and cesium in Par Pond. TID-26516 (NTIS), 5 p.

Du Pont de Nemours (E. I.) and Co., 1972: Environmental monitoring at the Savannah River plant, annual report, 1971. DPSPU-72-302 (NTIS), 29 p.

Du Pont de Nemours (E. I.) and Co., 1973: Environmental monitoring in the vicinity of the Savannah River plant, annual report for 1972. DPSPU-73-30-1 (NTIS), 27 p.

Du Pont de Nemours (E. I.) and Co., 1973: Environmental monitoring in the vicinity of the Savannah River plant, annual report for 1973. DPSPU-74-30-1 (NTIS), 31 p.

Du Pont de Nemours (E. I.) and Co., 1973: Environmental activities and programs at the Savannah River plant. DPST-73-436 (NTIS), 23 p.

Du Pont de Nemours (E. I.) and Co., 1974: Simple model to determine mesoscale transport of airborne pollutants. DP-MS-74-20 (NTIS), 25 p.

Harvey, R. S., Horton, J. H., and Mealing, H. G., 1959: Health physics regional monitoring semiannual report, January-June, 1958. DPSP-58-25-38.

Harvey, R. S., Horton, J. H., and Mealing H. G., 1959: Health physics regional monitoring semiannual report, July-December, 1958. DPSPU-59-11-23 (NTIS), 46 p.

Horton, J. H., and Mealing, H. G., 1956: Health physics regional monitoring semiannual report, June-December, 1955. DPSP-56-25-54(Del) (NTIS), 44 p.

Mealing, H. G., Jr., 1957: Health physics regional monitoring semiannual report, July-December, 1956. DPSP-57-25-15(Del), (NTIS), 38 p.

Mealing, H. G., Jr., and Horton, J. H., 1957: Health physics regional monitoring semiannual report, January-June, 1957. DPSP-57-25-43(Del), (NTIS), 43 p.

Mealing, H. G., Jr., and Harvey, R. S., 1958: Health physics regional monitoring semiannual report, July-December 1957. DPSP-58-25-17(Del), (NTIS), 50 p.

Reichert, S. O., 1962: Radionuclides in ground water at the Savannah River Plant waste disposal facilities. J. Geophys. Res., v. 67, n. 11, p. 4363-4374.

West Valley Plant

Nuclear Fuel Services, Inc., 1970: West Valley reprocessing plant, environmental report n. 7, July-December, 1969. DOCKET-50201-48 (NTIS), 12 p.

Nuclear Fuel Services, Inc., 1970: West Valley reprocessing plant, environmental report n. 8, January-June, 1970. DOCKET-50201-49 (NTIS), 12 p.

Nuclear Fuel Services, Inc., 1970: West Valley reprocessing plant, environmental report n. 9, July-December, 1970. DOCKET-50201-69 (NTIS), 12 p.

U.S. Atomic Energy Commission, 1970: West Valley reprocessing plant, environmental sample data. DOCKET-50201-62 (NTIS), 122 p.

B. SUBSURFACE NUCLEAR PROJECTS

General

Anspaugh, L. R., Koranda, J. J., and Robison, W. L., 1971: Environmental aspects of natural gas stimulation experiments with nuclear devices. Univ. California, Lawrence Radiation Lab., Rpt. UCRL-73429 (NTIS), 47 p.

Carter, M. W., and Moghissi, A. A., 1974: Radiation dose from natural gas vs that from nuclear stimulated natural gas (in 8th Midyear Topical Symposium of the Health Physics Society, October 21-24, 1974, Knoxville, Tenn.). CONF-741018, p. 355-360.

Green, J. B., Jr., and Lossler, R. M., 1971: Reduction of tritium from underground nuclear explosives. Univ. Californa, Lawrence Radiation Lab, Rpt. UCRL-73258 (NTIS), 27 p.

Ostlund, H. G., 1974: Study of tritiated hydrogen and its compounds in the environment, progress report, July 1, 1973 to March 31, 1974. ORO-3944-7 (NTIS), 24 p.

Retief, V., and Kruger, P., 1971: Use of nuclear explosives for water resources development in arid regions. Stanford Univ., Dept. Civil Engineering, Rpt. SU-326-P-31-2 (NTIS), 108 p.

Stead, F. W., 1963: Tritium distribution in ground water around large underground fusion explosions. Science, v. 142, p. 1163-1165.

Amchitka

Ballance, W. C., 1974: Radiochemical monitoring of water after the Cannikin event, Amchitka Island, Alaska August 1973. Amchitka-42 NTIS, USGS-474-205, 20 p.

Castagnola, D. C., 1969: Tritium anomalies on Amchitka Island, Alaska, part I. NTIS, NVO-1229, pt. 1, 41 p.

Castagnola, D. C., 1969: Tritium anomalies on Amchitka Island, Alaska, part II. NTIS, NVO-1229, pt. 2, 23 p.

Essington, E. H., Forslow, E. J., and Castagnola, D. C., 1970: Interim summary of radiochemical data for sts "A", Amchitka Island, Alaska through June 30, 1969. NTIS, NVO-1229-121, 62 p.

Essington, E. H., Forslow, E. J., and Castagnola, D. C., 1970: Interim summary of tritium data for sts "A", Amchitka Island, Alaska, July 1 1969 through June 30, 1970. NTIS, NVO-1229-157, 99 p.

Essington, E. H., 1971: An interim summary of tritium data for sts "A", Amchitka Island, Alaska, July 1, 1970 through June 30, 1971. NTIS, NVO-1229-172, 75 p.

Fenske, P. R., 1972: Event-related hydrology and radionuclide transport at the Connikin site, Amchitka Island, Alaska. NTIS, NVO-1253-1, 37 p.

Held, E. E., 1971: Amchitka radiological program, progress report, July 1970 through April 1971. NTIS, NVO-269-11, 38 pp.

Held, E. E., Nelson, V. A., Schell, W. R., and Seymour, A. H., 1973: Amchitka radiobiological program, progress report, March-December, 1972. NTIS, NVO-269-19, 97 p.

Merritt, M. L., 1971: Physical and biological effects, Milrow event, chapter 4-radioactivity. NTIS, NVO-79, p. 41-8.

Merritt, M. L., 1973: Physical and biological effects, Cannikin. NTIS, NVO-123, 106 p.

Nelson, V. A., and Seymour, A. H., 1974: Amchitka radiobiological program, progress report, January 1973 through December 1973. NTIS, NVO-269-1, 101 p.

Schroder, L. J., and Ballance, W. B., 1972: Radiochemical monitoring of water after the Cannikin event, Amchitka Island, Alaska, January 13 to April 5, 1972. NTIS, USGS-474-159 (Amchitka-35), 17 p.

<u>Nevada Test Site and Gasbuggy</u>

Clebsch, A., 1961: Tritium-age of ground water at the Nevada test site, Nye County, Nevada. U.S. Geol Survey, Prof. Paper 424-C, C.122-C.125.

Garber, M. S., 1971: Hydraulic-test and quality-of-water data from hole U-3cn PS#2, Bilby site, Nevada test site. U.S. Geol. Survey,

Kelly, M. J., Barton, C. J., Meyer, A. S., Chew, E. W., and Bowman, C. R., 1973: Experimental result from processing Gasbuggy gas in a natural gas processing plant (<u>in</u> Tritium). Phoenix, Arizona, Messenger Graphics, p. 603-611.

Mason, B. J., Hop, H. W., and Miller, C. L., 1973: Transfer of tritium from methane to vegetation (<u>in</u> Tritium). Phoenix, Arizona, Messenger Graphics, p. 455-461.

National Environmental Research Center, 1973: Environmental monitoring report for the Nevada test site and other test areas used for underground nuclear detonation, January-December 1972. NTIS, NERC-LV-539-23, 143 p.

National Environmental Research Center, 1974: Environmental monitoring report for the Nevada test site and other areas used for underground nuclear detonation, January-December 1973. NTIS, NERC-LV-539-31, 110 p.

National Environmental Research Center, 1974: Final report of the radiological surveillance program for the project Gasbuggy production test, May 15, 1973 to November 6, 1973. NTIS, NERC-LV-539-30, 63 p.

Public Health Service, 1970: Project Gasbuggy, off-site radiological safety report, Gb-2R phase I program. NTIS, SWRHL-105-R, 20 p.

Sokal, D., 1970: Ground water safety evaluation, project Gasbuggy. NTIS, PNE-1009, 38 p.

Southwestern Radiological Health Laboratory, 1970: Environmental surveillance for project Gasbuggy production, test phase. NTIS, SWRHL-100-r, 39 p.

Rio Blanco

Arnold, W. D., Salmon, R., Lin, K. H., and DeLaguna, W., 1973: Preliminary evaluation of methods for the disposal of tritiated water from nuclearly stimulated natural gas wells. NTIS, ORNL-TM-4024, 52 p.

Eberline Instrument Corp., 1971: Radiological monitoring program, period covering October 1, 1971 through December 31, 1971 for Cer geonuclear corporation, project Rio Blanco. NTIS, PNE-RB-51-1, 26 p.

Eberline Instrument Corp., 1972: Radiological monitoring program, period covering April 1, 1972 through June 30, 1972. NTIS, PNE-RB-51-3, 22 p.

Eberline Instrument Corp., 1973: Radiological monitoring program, period covering February 15, 1973 through May 16, 1973. NTIS, PNE-RB-51-5, 23 p.

Eberline Instrument Corp., 1973: Radiological monitoring program, period covering May 17, 1973 through June 31, 1973. NTIS, PNE-RB-51-6, 21 p.

Eberline Instrument Corp., 1973: Radiological monitoring program, period covering August 1, 1973 through October 31, 1973, for Cer geonuclear corporation, project Rio Blanco. NTIS, PNE-RB-51-7, 23 p.

Eberline Instrument Corp., 1974: Radiological monitoring program, period covering November 1, 1973 through February 4, 1974 for Cer geonuclear corporation, project Rio Blanco. NTIS, PNE-RB-51-8, 19 p.

Eberline Instrument Corp., 1974: Radiological monitoring program, period covering May 1, 1974 through July 29, 1974 for Cer geonuclear corporation, project Rio Blanco. NTIS, PNE-RB-5-10, 16 p.

Rulison

Claassen, H. C., and Voegeli, P. T., 1971: Radiochemical analyses of water samples from selected streams and precipitation collected in connection with calibration test flaring of gas from test well, August 15-October 13, 1970, project Rulison. Rulison-8, NTIS, USGS-474-107, 10 p.

Claassen, H. C., 1971: Radiochemical analyses of water samples from selected streams and precipitation collected October 23-November 9, 1970, in conjunction with the first production test flaring of gas from test well, project Rulison. Rulison-9, NTIS, USGS-474-108, 10 p.

Claassen, H. C., 1971: Radiochemical analyses of water from selected streams and precipitation collected immediately before and after the second production test flaring, project Rulison. NTIS, USGS - 474-122, 10 p.

Larson, J. D., and Beetem, W. A., 1970: Chemical and radiochemical analyses of water from streams, reservoirs, wells and springs in the Rulison project area, Garfield, and Mesa Counties, Colorado. Rulison-4, NTIS, USGS-474-67, 16 p.

Voegeli, P. T., Jr., 1971: Radiochemical analyses of water from selected streams, wells, springs, and precipitation collected prior to reentry drilling, project Rulison. Rulison-6, NTIS, USGS-474-83, 16 p.

Wrubel, D. T., Andrews, V. E., and Boysen, G. A., 1973: Environmental tritium surveillance for project Rulison (in Tritium). Phoenix, Arizona, Messenger Graphics, p. 592-602.

Salmon

Fenske, P. R., 1973: Hydrology and radionuclide transport monitoring well HT-2M, Tatum Dome, Mississippi. NTIS, NVO-1253-6, 19 p.

Janzer, V. J., Robinson, B. P., and Rucker, S. J., 1965: Radiochemical analysis of water samples collected after the Salmon event in the vicinity of Tatum Salt Dome, Lamar County, Mississippi. NTIS, USGS-474-115, 33 p.

Wagon Wheel

Eberline Instrument Corporation, 1973: Environmental monitoring report, period covering November 1, 1973-January 31, 1973. NTIS, PNE-WW-28, 26 p.

Eberline Instrument Corporation, 1973: Environmental monitoring report, period covering February 1, 1972-April 30, 1973. NTIS, PNE-WW-30, 24 p.

Eberline Instrument Corporation, 1973: Environmental monitoring report, period covering May 1, 1973-July 31, 1973. NTIS, PNE-WW-32, 25 p.

Eberline Instrument Corporation, 1973: Environmental monitoring report, period covering August 1, 1973-December 8, 1973 for El Paso Natural Gas Company, project Wagon Wheel. NTIS, PNE-WW-34, 26 p.

C. REACTORS

General

de Bortoli, M., and Gaglioni, P., 1974: Environmental radioactivity, Ispra, 1972. Commission of the European Communities, Joint Nuclear Research Center, Rpt. EUR-5118 (NTIS), 63 p.

Draley, J. E., and Greenberg, S., 1972: Some features of the impact of a fusion reactor power plant on the environment. Argonne National Laboratory Rpt. CONF-721111-22 (NTIS), 38 p.

Elwood, J. W., 1971: Ecological aspects of tritium behavior in the environment. Nuclear Safety, v. 12, n. 4, p. 325-337.

Gavin, A. P., Breden, C. R., Bailey, R. E., Theys, M. H., and Goring, G. E., 1959. Water decomposition in a direct cycle boiling water reactor. Indust. Eng. Chem., v. 51, p. 1265-1266.

Grathwohl, G., 1972: Production and release of tritium by nuclear power plants and reprocessing plants and the expected radiological burden till the year 2000 (German). Institut fur Neutronenphysik und Reaktor Technik, Rpt. NP-19896, 83 p.

Gupta, V. K., and Soman, S. D., 1973: Environmental tritium contamination from nuclear power program (<u>in</u> Tritium). Phoenix, Arizona, Messenger Graphics, p. 564-571.

Krieger, H. L., Gold, S., and Kahn, B., 1973: Tritium releases from nuclear power stations (<u>in</u> Tritium). Phoenix, Arizona, Messenger Graphics, p. 557-564.

Musgrave, B. C., 1974: Tritium distribution in the nuclear industry--the requirement for control strategies. Allied Chemical Corp., Idaho Chemical Program Operation Office, Rpt. ICP-1041, 23 p.

Osborne, R. V., 1971: Monitoring reactor effluents for tritium--problems and possibilities. NTIS, AECL-4054, 31 p.

Porter, S. W., Jr., 1973: The monitoring of tritium in the aquatic environment of power reactors (in Tritium). Phoenix, Arizona, Messenger Graphics, p. 517-521.

Soman, S. D., and Abraham, P., 1967: Estimation of dilution rate factors for tritiated water vapour from a reactor stack. Health Physics, v. 13, n. 10, p. 1117-1121.

Steele, T. A., 1973: Tritium generation and release to in-plant and off-site environs of the LaCrosse boiling water reactor (in Tritium). Phoenix, Arizona, Messenger Graphics, p. 585-591.

Weaver, C. L., Harward, E. D., and Peterson, H. T., Jr., 1969: Tritium in the environment from nuclear power plants. U.S. Dept. Health, Education and Welfare, Public Health Rpts., v. 84, n. 4, p. 363-371.

Specific

Barraclough, J. T., Teasdale, W. E., and Robertson, J. B., 1967: Hydrology of the national reactor testing station, Idaho, 1966. U.S. Geol. Survey Rpt. IDO-22049 (NTIS), 95 p.

Cohen, L. K., and Kneip, T. J., 1973: Environmental tritium studies at a power plant (pwr), (in Tritium). Phoenix, Arizona, Messenger Graphics, p. 623-639.

Ehhalt, D. H., and Bainbridge, A. E., 1966: A peak in the tritium content of the atmospheric hydrogen following the accident at Windscale. Nature, v. 209, n. 5026, p. 903-904.

Eliason, J. R., 1967: Field evaluation of ground disposal of reactor effluent, 1301-N-Crib. Battelle Pacific Northwest Laboratories Rpt. BNWL-CC-1032 (NTIS), 22 p.

Hawkins, D. B., and Schmalz, B. L., 1967: Environmental tritium studies at the national reactor testing station (in Isotope Techniques in the Hydrologic Cycle). Am. Geophys. Union, Geophysical Mon. Ser. n. 11, p. 157-158.

Kahn, B., Blanchard, R. L., Kolde, H. E., Krieger, H. L., and Gold, S., 1971: Radiological studies at a pressurized water nuclear power reactor. National Environmental Research Center Rpt. EPA-RD-71-1, 105 p.

Yankee Power Reactor

Kirchmann, R., and Cantillon, G., 1971: Comportement en rivière des radiocobalts et du radiomanganese provenant d'effluents d'une centrale pwr (Behavior in rivers of radiocobalt and radiomanganese in effluents from pwr nuclear power plants) (in European Communities International Symposium, Radioecology Applied to the Protection of Man and his Environment, September 1971, Rome). 17 p.

Mühlhölzl, W. Z., 1969: Tritium determination in reactor water, wastewaters, and receiving waters from the nuclear power plant Gundremmingen/Danube in 1968. Wass. Abwass. Forsch., v. 2, p. 222-226.

Northern States Power Co., 1974: Monticello nuclear generating plant, unit 1, semiannual operating report no. 7, January-June, 1974. NTIS, DOCKET-50263-413, 118 p.

Omaha Public Power District, 1973: Fort Calhoun station, unit 1, environmental radioactivity surveillance program pre-operational report, second quarter, 1973. NTIS, DOCKET-50285-163, 35 p.

Robertson, J. B., and Barraclough, J. T., 1973: Radioactive and chemical waste transport in ground water at national reactor testing station, Idaho; 20-year case history and digital model (abstr.). Am. Assoc. Pet. Geol., Bull., v. 57, n. 8, p. 1603-1604.

Robertson, J. B., Schoen, R., and Barraclough, J. T., 1974: Influence of liquid waste disposal on the geochemistry of water at the national reactor testing station, Idaho, 1952-1970. NTIS, IDO-22053, 231 p.

Service Central de Protection Contre les Rayonnements Ionisants, 1973: Report of activity, February 1973. SCPRI(RM)-2-1973 (NTIS), 12 p.

Service Central de Protection Contre les Rayonnements Ionisants, 1973: Report of activity, April 1973: SCPRI(RM)-4-1973 (NTIS), 12 p.

Service Central de Protection Contre les Rayonnements Ionisants, 1973: Annex to monthly report, May 1973. Detail of results of measurement. SCPRI(RM)-5-1973 (suppl.), (NTIS), 23 p.

Service Central de Protection Contre les Rayonnements Ionisants, 1973: Monthly progress report (June). SCPRI(RM)-6-1973 (NTIS), 19 p.

Service Central de Protection Contre les Rayonnements Ionisants, 1973: Details of results of measurements. SCPRI(RM)-6-1973 (suppl.), (NTIS), 30 p.

Service Central de Protection Contre les Rayonnements Ionisants, 1973: Monthly progress report (July). SCPRI(RM)-7-1973 (NTIS), 15 p.

Service Central de Protection Contre les Rayonnements Ionisants, 1973: Appendix to the monthly progress rep report, details to the report of measurements. SCPRI-(RM)-7-1973 (suppl.), (NTIS), 25 p.

Service Central de Protection Contre les Rayonnements Ionisants, 1973: Monthly progress report (August). SCPRI(RM)-8-1973 (NTIS), 13 p.

Service Central de Protection Contre les Rayonnements Ionisants, 1973: Monthly progress report (September). SCPRI(RM)-9-1973 (NTIS), 15 p.

Service Central de Protection Contre les Rayonnements Ionisants, 1973: Report of activity, October 1973. SCPRI(RM)-10-1973 (NTIS), 18 p.

Service Central de Protection Contre les Rayonnements Ionisants, 1973: Report of activity, November 1973. SCPRI(RM)-11-1973 (NTIS), 15 p.

Service Central de Protection Contre les Rayonnements Ionisants, 1973: Annex to monthly report, November 1973, detail of results of measurement. SCPRI(RM)-11-1973 (suppl.), (NTIS), 22 p.

Service Central de Protection Contre les Rayonnements Ionisants, 1973: Report of activity, December 1973. SCPRI(RM)-12-1973 (NTIS), 16 p.

Service Central de Protection Contre les Rayonnements Ionisants, 1973: Annex to monthly report, December 1973, detail of results of measurements. SCPRI(RM)-12-1973 (suppl.), (NTIS), 27 p.

Service Central de Protection Contre les Rayonnements Ionisants, 1974: Report of activity, January 1974. SCPRI(RM)-1-1974 (NTIS), 15 p.

Service Central de Protection Contre les Rayonnements Ionisants, 1974: Annex to monthly report, January 1974, detail of results of measurements. SCPRI(RM)-1-1974 (suppl.), (NTIS), 23 p.

Service Central de Protection Contre les Rayonnements Ionisants, 1974: Report of activity, February 1974. SCPRI(RM)-2-1974 (NTIS), 15 p.

Service Central de Protection Contre les Rayonnements Ionisants, 1974: Annex to monthly report, February 1974, detail of results of measurements. SCPRI(RM)-2-1974 (suppl.), (NTIS), 23 p.

D. NUCLEAR WASTE DISPOSAL

Haney, W. A., Brown, D. J., and Reisenauer, A. E., 1962: Fission-product tritium in separation wastes and in the ground water. U.S. Atomic Energy Comm., HW-74536, 14 p.

Kolychev, B. S., Kulichenko, V. V., and Rauzen, F. V., 1972: The status of the radioactive waste disposal problem. Soviet Atomic Energy, v. 33, n. 3, p. 837-841.

Leveque, P. Ch., Giannotti, C. P., Grison, G., et al., 1971: Critères hydrodynanimiques de garantie de pérennite pour le stockage géologique supérficiel des déchets radioactifs (Hydrodynamic criteria for assuring the perpetuity of superficial geologic disposal of radioactive wastes) (in Congress of Tokyo, Asian Regional Conference, Réunion de Tokyo, Conference Régionale de l'Asie). Assoc. Internat. Hydrogeol. Mem., v. 9, p. 145-153.

Magno, P., Reavey, T., and Apidianakis, J., 1970: Liquid waste effluents from a nuclear fuel reprocessing plant. Radiological Health Lab., Rpt. BRH/NERHL-70-2 (NTIS), 69 p.

14. SUPPORTIVE TECHNIQUES

Elrick, D. E., and Lawson, D. W., 1969: Tracer techniques in hydrology (in Hydrology Symposium, 7th, October 8-9, 1969, Victoria, B.C., Instrumentation and Observations Techniques). Victor, B.C., v. 1, p. 155-187.

Fireman, E. L., D'Amico, J., DeFelice, J., et al., 1972: Radioactivities in returned lunar materials (in Lunar Science Conference, 3rd proceedings). Geochim. Cosmochim. Acta, Suppl., v. 2, n. 3, p. 1747-1761.

Huff, D. D., and Kruger, P., 1967: Numerical model for the hydrologic transport of radioactive aerosols from precipitation to waste supplies. Am. Geophys. Union, Geophysical Mono. Ser., n. 11, p. 85-96.

Makowski, J., 1972: Die wahl von radioisotopen und deren aktivitaet auf grund von modellmessungen (The appropriate choice of radioisotopes and their activities for hydrogeological studies based on model measurements) (in Internationale fachtagung zur untersuchung unterirdischer wasserwege mittels kuenstticher und natürelicker markierungsmittel, 2nd, vortraege diskussionen und beitraege, ger., bundesant) with English, French and Russian summaries. Bodenforsch. Geol. Landesaemt., Geol. Jahrb., Reihe C, n. 2, p. 329-337.

Parker, F. L., 1958: Radioactive tracers in hydrologic studies. Am. Geophys. Union, Trans., v. 39, n. 3, p. 434-439.

Massachusetts study

Payne, B. R., 1967: Isotope techniques in ground water hydrology (<u>in</u> Methods and Techniques of Ground Water Investigation and Development, U.N. Econ. Comm. Asia-Far East). Water Resources Res., n. 33, p. 107-114.

Rákóczi, L., 1966: The use of radioactive isotopes in hydraulic and hydrological research. VITUKI, 166 p.

Suska, J., 1958: O izotopach radioaktywnych (Radioactive isotopes). Gazela obserwatora PIHM, v. 11, n. 8, p. 3-7.

Vartazarov, S. I., et al., 1968: Radioactive tracers in hydrological researches in U.S.S.R. Internat. Assoc. Sci. Hydrol., Publ. n. 78, p. 156-161.

Vilenskiy, V. D., 1972: Radioaktiunyye izotopy v lednikovom pokrove Antarktidy (Radioactive isotopes in the Antarctic ice sheet). Antarktika, n. 11, p. 157-173.

PART II

DEUTERIUM AND OXYGEN

1. BIBLIOGRAPHY

International Atomic Energy Agency, 1973: Isotope techniques in hydrology, v. 11 (1966-1971). IAEA Bibliogr. Ser. n. 41, 233 p.

814 references

Samuel, D., and Steckel, F., 1959: Bibliography of the stable isotopes of oxygen (O^{17} and O^{18}). Pergamon Press, New York, 224 p.

725 references

Weed, H. C., 1972: Reactions of methane and of carbon monoxide with tritiated water, a literature review. Lawrence Livermore Laboratory, Univ. California (also NTIS), 16 p.

Hoering, T. C., 1960: Stable-isotope geochemistry. Am. Geophys. Union, Trans., v. 41, n. 2, p. 298.

29 references

Tilton, G. R., 1971: Radioactive and radiogenic isotope research (<u>in</u> Volcanology, Geochemistry and Petrology, U.S. National Report, 1967-1971, 15th General Assembly, IUGG). EOS, v. 52, n. 5, p. IUGG 100-105.

Weber, J. N., 1964: Bibliography--geochemistry of the stable isotopes of carbon and oxygen. Penn. State Univ., Mineral Industries Expt. Sta. Circ. 67, 80 p.

2. SYMPOSIA, CONFERENCES AND COLLECTIONS

Czachorski, W., (ed), 1969: Applications of isotopes in geophysics, hydrogeology and engineering geology (Engl. trans). Nucleonika, v. 13, n. 4-5, 224 p.

Dontsova, Ye. I., 1968: Isotopen geologie, bericht über ein symposium in der UdSSR (Isotope geology, report on a symposium in the USSR). Z. Angew Geol., v. 14, n. 10, p. 541-544.

Dontsova, Ye. I., 1973: IV Vsesoyuznyy simpozium po primeneniyu stail'nykh izotopov v geokhimii (The fourth all-union symposium on the application of stable isotopes in geochemistry). Geokhim Akad. Nauk SSSR, n. 4, p. 947-951.

Dostal, P., and May, F., 1969: Zur anwendung stabiler isotope in der geochemie (On the application of stable isotopes in geochemistry). Zeitschr. Angew. Geologie, v. 15, n. 10, p. 527-530.

Report on Moscow Symposium, 1968

Dowgiallo, J., 1974: Miedzynarodowe sympozjum na temat stosowania technik izotopowych w hydrogeologii (International symposium on the use of isotope techniques in hydrogeology). Przegl. Geol., v. 22, n. 10, p. 517-518.

International Atomic Energy Agency, 1970: Isotope hydrology proceedings of symposium on the use of isotope techniques in hydrology, 1970. Vienna, IAEA, STI-Pub. 255, 918 p.

54 papers

International Symposium on Atmospheric Trace Constituents and Atmospheric Circulation, 1970: Proceedings, J. Geophys. Res., v. 75, n. 15, p. 2875.

Kitona, Y. (ed), 1975: Geochemistry of water (in Benchmark Papers in Geology). Dowden, Hutchinson and Ross, Inc., 455 p.

Contains 21 papers previously published

Maass, I., and Geissler, C., 1970: Arbeitstagung über stabile isotope 12. bis 19. Oktober 1979 in Leipzig (Workshop on stable isotope, October 12 to 19, 1969 in Leipzig). Zeitschr. Angew. Geologie, v. 16, n. 11-12, p. 489-490.

Renaud, A., et al., 1969: Etudes physiques et chimiques sur la glace de l'Indlandsis du Groenland 1959 (Physical and chemical studies on the ice of the Greenland ice cap 1959), English and German Abstracts. Exped. Glacial. Internat. Groenland, v. 5, n. 3; Medd. Grönland, v. 177, n. 2, 123 p.

Stout, G. E., 1967: Isotope techniques in the hydrologic cycle symposium, 1965, Univ. Illinois. Am. Geophys. Union, Geophysical Mon. Ser. n. 11, 199 p.

3. HISTORY AND STATE OF THE ART

Bigeleisen, J., 1965: Chemistry of isotopes. Science, v. 147, n. 3657, p. 463-471.

Sulfur and oxygen

Brill, R. H., 1970: Lead and oxygen isotopes in ancient objects (in A Symposium on the Impact of the Natural Sciences on Archeology). Royal Soc. London Philos. Trans., Ser. A, v. 269, n. 1193, p. 143-164.

Clayton, R. N., 1971: Stable isotope geochemistry (in Volcanology, Geochemistry and Petrology, U.S. National Report, 1967-1971, 15th General Assembly, IYGG). EOS, v. 52, n. 5, p. IUGG 106-110.

Dontsova, Ye. I., 1967: Informatisiya ob issledovaniyakh v oblasti izotopnoy geologii (Information on investigations in the field of isotope geology). Geokhimiya, n. 5, p. 641-646.

Dontsova, Ye. I., and Grinenko, V. A., 1971: Geokhimiya stabil'nykh izotopov (Geochemistry of the stable isotopes) (in Problemy geologii na XXIII sessii mezhdunarodnogo geologicheskogo kongressa). Moscow, Akad. Nauk SSSR, Nats. Kom. Geol. Sov. Soyuza, p. 105-108.

Epstein, S., 1959: The variations of O^{18}/O^{16} ratio in nature and some geologic implications (in Abelson, P. H., ed., Researches in Geochemistry). New York, John Wiley, p. 217-240.

Folinsbee, R. E., 1967: Isotope studies (in Canadian Upper Mantle Report, 1967). Canada Geol. Surv. Paper 67-41, p. 147-154.

Friedman, I., 1962: Recent applications of deuterium abundance measurements to the earth sciences (abstr). J. Geophys. Res., v. 67, n. 4, p. 1637.

Friedman, I., 1962: Stable isotopes. Am. Geophys. Union, Trans., v. 144, n. 2, p. 521-523.

Hoefs, J., 1973: Stable Isotope Geochemistry. New York, Springer-Verlag, Inc., 151 p.

Ingerson, E., 1953: Non-radiogenic isotopes in geology, a review. Geol. Soc. Am. Bull., v. 64, n. 3, p. 301-374. Fractionation

Jensen, M. L., 1953: The geologic importance of variations in stable isotopic abundances. Econ. Geology, v. 48, n. 3, p. 161-173.

Krüger, P., 1968: Zur gegenstandsbestimmung der isotopengeochemie; ein beitrag zur klärung der beziehungen zwischen radiogeologie, isotopengeologie und isotopengeochemie (The scope of isotope geochemistry; a contribution to clarification of the relationship between radiogeology, isotope geology, and isotope geochemistry). Deut. Ges. Geol. Wiss., Ber., Reihe A, Geol. Palaontol., v. 13, n. 1, p. 121-126.

Leutwein, F., 1968: Geochemie (Geochemistry), (in Vom erdkern bis zur magnetosphäre aktuelle probleme der erdwissenschaften). Frankfort-am-Main, Umschau, Verlag, p. 173-190.

Panichi, C., and Vento, D., 1971: Cenni sull'utilizzazione degli isotopi stabili dell'assigeno e dell'idrogeno per gli studi meteorologici (in Some Remarks on Meteorological Applications of Stable Oxygen and Hydrogen Isotopes), English and French summaries. Rome, Rivista di Meteorologia Aeronautica, v. 31, n. 4, p. 377-385.

Perry, E. C., Jr., 1969: Oxygen isotope studies (in Summary of Fieldwork). Minn. Geol. Surv. Inform. Circ. n. 7, p. 24.

Perry, E. C., Jr., 1970: Stable isotope studies (in Summary of Fieldwork). Minn. Geol. Surv. Inform. Circ. n. 8, p. 15.

Rankama, K., 1954: Isotope geology. London, Pergamon Press, 535 p.

Rankama, K., 1958: Aplicaciones geológicas de la radioactividad y de los nucleidos estables (Geologic applications of radioactivity and the stable nuclides), (in Cursillos y conferencias del Instituto "Lucas Mallada," no. 5). Inst. Lucas Mallada Inv. Geol., p. 3-15.

Carbon, oxygen and sulfur

Rankama, K., 1963: Progress in isotope geology. New York-London, Interscience Pub. (John Wiley and Sons), 705 p.

Rösler, H.-J., and Pilot, J., 1969: Anwendung und bedeutung der isotopengeochem (Application and significance of isotope geochemistry), English and Russian summaries. Zeitschr. Angew. Geologie, v. 15, n. 9, p. 491-500.

Schatenstein, A. I., Jakowlewa, E. A., Swjaginzewa, E. N., Warschawski, Ja. M., Israilewitsch, E. A., and Dychno, N. M., 1960: Isotopenanalyse des wassers (Isotope analysis of water). Berlin, VEB Deutscher Verlag der Wissenschaften, 270 p.

Sheppard, P. A., 1957: Natural and artificial trace elements in geophysical research. London, Nature, v. 179, n. 4568, p. 996-997.

Silverman, S. R., 1951: The isotope geology of oxygen. Geochim. Cosmochim. Acta, v. 2, n. 1, p. 26-42.

Taylor, H. P., Jr., 1966: Isotope geochemistry of oxygen and hydrogen-a brief review. Am. Geophys. Union, Trans., v. 47, n. 1, p. 287-290.

Taylor, H. P., Jr., 1967: Stable isotopes (in IUGG Quadrennial Report, U.S.A.). Am. Geophys. Union, Trans., v. 48, n. 2, p. 686-693.

Thode, H. G., 1964: Stable isotopes--a key to our understanding of natural processes. Bull. Can. Petroleum Geology, v. 12, n. 2, p. 246-262.

Vinogradov, A. P., 1954: Geokhimiya isotopov (The geochemistry of isotopes). Akad. Nauk SSSR Izv. Ser. Geol., n. 3, p. 3-19.

Carbon, oxygen, sulfur and hydrogen

Viswanathon, S., 1972: How old is silicate oxygen isotope geochemistry? Who were the pioneers? Geol. Mag., v. 109, n. 3, p. 277-280.

Volborth, A., 1968: On the distribution and role of oxygen in the geochemistry of the earth's crust (in Ahrens, L.H., ed., Origin and Distribution of the Elements). Oxford, England, New York, Pergamon Press, Internat. Ser. Mons. Earth Sci., v. 30, p. 825-853.

Wallo, E. M., and Remson, I., 1963: Applications of tracers in studies of soil water movement. Am. Geophys. Union, Trans., v. 44, n. 2, p. 576-577.

Wanless, R. K., 1957: Application of isotopic studies to geological problems. Canadian Mining J., v. 78, n. 4, p. 133-136.

4. PROPOSALS FOR USE

Chilingar, G. V., 1955: O^{18}/O^{16} ratio on carbonate rock an accurate geologic thermometer? Brief review of Russian literature. Am. Assoc. Petrol. Geologists Bull., v. 39, n. 11, p. 2349-2350.

Degens, E. T., 1960: Stabile isotopes in ihren beziehungen zur zeit (Stable isotopes in their relations to time). Geol. Rundschau, v. 49, n. 1, p. 277-278.

Ferronskiy, V. I., 1967: Prirodnyye izotopy i ikh ispol' zovaniye pri izuchenii gidrogeologicheskikh protsessov (Natural isotopes and their use in the study of hydrogeologic processes). Mosk Obshchest. Ispyt. Prir. Otd. Geol., v. 42, n. 4, p. 111-115.

Jensen, M. L., 1970: Stable isotopes in geochemical prospecting abs. (in International Geochemical Exploration Symposium, 3rd, Program and Abstracts). Can. Inst. Mining Met. Geol. Div., Soc. Econ. Geol., p. 40.

Jensen, M. L., 1971: Stable isotopes in geochemical prospecting (in Geochemical Exploration, International Geochemical Exploration Symposium, 3rd Proceedings). Can. Inst. Min. Met., Spec. Vol., n. 11, p. 464-468.

Jensen, M. L., 1973: Mineral exploration and stable isotopes (abs.). Geol. Soc. Am., Abs., v. 5, n. 7, p.682-683.

Münnich, K. O., 1969: Anwendung von isotopenanalysen an sauerstoff und kohlenstoff in der geologie (Geologic applications of oxygen and carbon isotope analyses). Deut. Geol. Ges. Z., v. 118, pt. 2, p. 217.

Nestler, P., 1974: Use of the oxygen isotopic analysis in the solution of petrological and geochemical problems. Z. Geol. Wiss., v. 2, n. 6, p. 691-703.

Pilot, J., and Rosler, H. J., 1968: Isotopenparagenese als hilfsmittel geologisch--geochemischer aussagen (Isotope paragenesis as an aid to geological-geochemical statements). Freiberger Forschungshefte, C 231, p. 267-275.

Polánsky, A., 1968: Anwendungsmöglichkeiten isotopengeochemischer untersuchungen zur lösung geologischer probleme (Possibilities of applying isotope geochemical investigations to the solution of geologic problems). Deutsch. Gesell. Geol. Wiss. Ber., ser. B, v. 13, n. 2, p. 237-249.

Sulfur, oxygen and carbon

Povarennykh, A. S., 1966: Rozvytok novykh napryamkiv i metodiv u vychenni krystalichnoyi rechovyny zemnoyi v kory (Development of new directions and methods in the study of the crystalline matter of the earth's crust). Akad. Nauk Ukrayin. RSR Heol. Zhur., v. 26, p. 69-74.

5. METHODS FOR MEASUREMENT

Beckinsale, R. D., Durham, J. J., Freeman, N. J., et al., 1973: Some recent progress in high precision determination of O^{18} abundances in geological material, abs. (in Europacisches Kolloquium fuer Geochronologie, II). Referate, Fortschr. Mineral., v. 50, Beih. 3, p. 49.

Berezovskiy, F. I., Demidenko, S. G., and Lugovaya, I. P., 1967: Metod vyvchennya izotopnoho skladu kysnyu zalizynykh rud ta pryrodny kh vod (A method of studying the isotopic composition of oxygen in iron ores and natural waters), with English and Russian summaries. Akad. Nauk Ukrayin. RSR Dopovidi, Ser. B, n. 7, p. 642-645.

Blattner, P., 1973: Oxygen from liquids for isotopic analysis and a new determination of αCO_2-H_2O at 25°C. Geochim. Cosmochim. Acta, v. 37, n. 12, p. 2691-2693.

Borgest, V. A., 1961: Determination of deuterium in water at concentrations close to the natural abundance on the basis of infra-red absorption spectra. Opt. i. Spektroskop., v. 11, p. 558-559.

Borshchevskiy, Yu. A., Amosova, Kh. B., Borisova, S. L., Ustinov, V. I., and Savchenko, Yu. I., 1971: Novyy metod vydeleniya elementov iz mineralov dlya izotopnogo analiza (A new method of separating elements from minerals for isotope analysis). Akad. Nauk SSSR Doklady, v. 196, n. 5, p. 1203-1205.

Boyer, P. D., Graves, D. J., Suelter, C. H., and Dempsey, M. E., 1961: Simple procedure for conversion of oxygen of orthophosphate or water to carbon dioxide for oxygen-18 determinations. Anal. Chem., v. 33, p. 1906-1909.

Bruner, F., Ciccioli, P., and DiCorcia, A., 1972: Gas chromatographic analysis of complex deuterated and tritiated mixtures with packed columns. Anal. Chem., v. 44, n. 6, p. 894-898.

Cannavale, L., 1960: Determination of low concentration of D_2O in water by the flotation method. Buenos Aires, Com. Nac. Energ. Atom, Rpt. 27, 88 p.

Clayton, R. N., and Mayeda, T. K., 1963: The use of bromine pentafluorine in the extration of oxygen from oxides and silicates for isotopic analysis. Geochem. Cosmochim. Acta, v. 27, n. 1, p. 43-52.

Coplen, T. B., and Clayton, R. N., 1973: Hydrogen isotopic composition of NBS and IAEA stable isotope water reference samples. Geochim. Cosmochim. Acta, v. 37, n. 10, p. 2347-2349.

Craig, H., 1957: Isotopic standard for carbon and oxygen and correction factors for mass spectrometric analysis of carbon dioxide. Geochim. Cosmochim. Acta, v. 12, n. 1-2, p. 133-149.

Craig, H., 1961: Standard for reporting concentrations of deuterium and oxygen-18 in natural waters. Science, v. 133, p. 1833-1834.

Dattner, J., and Fischler, J., 1963: Calculations of oxygen isotope abundance ratios in a mass spectrometer. British J. Appl. Physics, v. 14, n. 10, p. 728-729.

Dontsova, Ye. I., 1969: Metod opredeleniya sootnosheniy izotopov kisloroda v gornykh porodakh i mineralakh (Methods of determining the oxygen isotope ratios in rocks and minerals), English trans. Geokhimiya, n. 8, p. 824-838.

Friedman, I., and Gleason, J. D., 1973: A new silicate intercomparison standard for O^{18} analysis. Earth Planet. Sci. Lett., v. 18, n. 1, p. 124.

Friedman, I., and Gleason, J. D., 1973: Notes on the bromine, pentafluoride technique of oxygen extraction. J. Res. U.S. Geol. Surv., v. 1, n. 6, p. 679-680.

Grosse, A. V., Johnston, W. M., Wolfgang, R. L., and Libby, W. F., 1951: Tritium in nature. Science, v. 113, p. 1-2.

Jackson, H. G., Libby, L. M., and Lukens, H. R., 1973: Measurements of O^{18}/O^{16} ratio using a fast neutron reactor. J. Geophys. Res., v. 78, n. 30, p. 7145-7148.

Jackson, M. C., and Young, W. A. P., 1973: Capacitive integration system for the precise measurement of isotopic ratios in a mass spectrometer. N.Y. Rev. Sci. Instr., v. 44, n. 1, p. 32-34.

Javoy, M., Fourcade, S., and Allegre, C. J., 1970: Graphical method for examination of O^{18}/O^{16} fractionations in silicate rocks. Earth Planet. Sci. Lett., v. 10, n. 1, p. 12-16.

Keder, W. F., and Kalkwarf, D. R., 1966: Near infra-red determination of deuterium oxide in water. Anal. Chem. v. 38, p. 1288.

Khaytov, B. K., 1969: Izotopy v gidrogeologii (Isotopes in hydrogeology). Akad. Nauk Uzb. SSR, 149 p.

Kirshfeldt, Yu. E., 1964: Mass-spektrometricheskiy analiz i primeneniye yego v neftyanoy geologii (Mass spectrometric analysis and its use in petroleum geology) (in Seysmicheskiya, Gravimetricheskiye i Promyslovo-geofizicheskiye, Issledovaniya v Neftyanov i Gazovoy Promyshlennosti). Moskov. Inst. Neftekhim. i Gazov. Promyshlenosti Trudy, n. 50, p. 224-231.

Krishna-Murthy, P. V., and Prahallada-Rao, B. S., 1963: An all-metal spectrometer for the isotopic analysis of hydrogen at natural concentration levels. Indian J. Pure Appl. Physics, v. 1, n. 2, p. 73-79.

Létolle, R., Marce, A., and Fontes, J.-C., 1965: Un spectromètre de masse pour mesure de précision des abondances isotopiques de l'oxygène et du carbone (A mass spectrometer for precision measurement of the isotopic abundance of oxygen and carbon). Soc. Francaise Mineralogie et Cristallographie Bull., v. 88, n. 3, p. 417-421.

Lugova, I. P., 1973: Metod vidilennya kishyu z karbonativ zaliza z metoya izotopnogo analizu (Method of oxygen extraction from iron carbonates for isotopic analysis), English and Russian summaries. Akad. Nauk Ukr. RSR, Dopov., Ser. B, n. 4, p. 304-308.

Maass, I., 1962: Beitrage zur isotopen geologie an den elementen wasserstoff,kohlenstoff und sauerstoff (Contributions to isotope geology on the elements hydrogen, carbon and oxygen). Geol. Gesell. DDR Ber., v. 6, n. 4, p. 408-418.

Matsubaya, O., 1972: The effect of pH of water on the measurement of oxygen isotopic ratio by means of CO_2-H_2O isotopic exchange technique. English and Japanese summaries. Okayama Univ. Inst. Therm. Spring Res., Paper n. 41, p. 1-2.

Mayne, K. I., 1960: Stable isotope geochemistry and mass spectrometric analysis (in Methods in Geochemistry). New York, London, Interscience (John Wiley and Sons), p. 148-201.

McCarthy, J. H., et al., 1961: Density comparison method for the determination of O^{18}/O^{16} ratios in prepared waters. U.S. Geol. Survey, Paper 424, p. 387-389.

Mizutani, Y., 1971: An improvement in the carbon-reduction method for the oxygen isotopic analysis of sulphates. Geochem. J. (Geochem. Soc. Jap.), v. 5, n. 2, p. 69-77.

O'Neil, J. R., and Epstein, S., 1966: A method for oxygen isotope analysis of milligram quantities of water and some of its applications. J. Geophys. Res., v. 71, n. 20, p. 4955-4961.

Pearson, R. T., and Derbyshire, W., 1974: NMR studies of water adsorbed on a number of silica surfaces. J. Colloid and Interface Sci., v. 46, n. 2, p. 232-248.

Sharma, T., and Clayton, R. N., 1965: Measurement of O^{18}/O^{16} ratio of total oxygen of carbonates. Geochim. Cosmochim. Acta, v. 29, n. 12, p. 1347-1353.

Thurston, W. M., 1970: Automatic mass spectrometric analysis of the deuterium to hydrogen ratio in natural water. Rev. Sci. Instr., v. 41, n. 7, p. 963-966.

Thurston, W. M., 1971: Steam film sampling of water for mass spectrometric analysis of the deuterium content. Rev. Sci. Instr., v. 42, n. 5, p. 700-703.

Trimborn, P., 1971: Massenspektrometrische messungen des O^{18}-gehalt von wassenproben. Universitat München.

Vetshteyn, V. Yu., Abashidze, I. V., Artemchuk, V. G., et al. : Pretsiziyniy mas-spektrometr dlya viznachennya vmistu deyteriyu prirodnikh spolukahn (Precision of the mass spectrometer for determining deuterium in natural compounds), English summary. Akad. Nauk Ukr. RSR, Dopov. Ser. B, n. 3, p. 198-200.

Vetshteyn, V. Yu., Miroshnichenko, A. G., and Artemchuk, V. G., 1972: Metod pretsiziynogo ekspres-analizu izotopnogo sklada vodnyu prirodnikh vod (A quick method of precise analysis of isotopic composition of hydrogen from natural waters), English and Russian summaries. Akad. Nauk Ukr. RSR, Dopov. Ser. B, n. 7, p. 632-634.

Watanabe, M., and Matsubaya, O., 1972: On the measurement of oxygen and carbon isotopic ratios of carbonates. English and Japanese summaries. Okayama Univ. Inst. Therm. Spring Res., Paper n. 41, p. 9-12.

6. PHYSICS, CHEMISTRY AND THERMODYNAMICS

Broadwater, T. L., and Evans, D. F., 1974: The conductance of divalent ions in H_2O at 10 and 25°C and in D_2O. J. Solution Chem., v. 3, n. 10.

Chong, S., et al., 1971: The proton affinity of water. Purdue Univ. Project Squid Technical Rpt. (Rice-11-Pu), 10 p.

Clark, S. P., Jr., 1966: Isotopic abundances and 1961 atomic weights, sec. 3 (in Handbook of Physical Constants, rev. edition). Geol. Soc. America Mem. 97, p. 11-17.

Goudeket, M., 1951: Isothermen van watersof und deuterium; tussen 0°C tot drukken van 3000 atmosferen (Isotherms of hydrogen and deuterium between 0°C and 150°C with a pressure of 3000 atmospheres). Asseh, Van Gorcum, 74 p.

Kudish, A. I., Wolf, D., and Steckel, F., 1974: Physical properties of oxygen-17 water; absolute viscosity and density of H_2O^{17} between 15 and 35°C. J. Chem. Soc. Faraday Trans. I, n. 3, p. 484-489.

McLaughlin, E., 1973: Isotope effect and the molecular mechanism of the second viscosity coefficient of water. J. Phys. Chem., v. 77, n. 14, p. 1801-1802.

Pupezin, J., Jakli, G., Jancso, G., and Van Hook, W. A., 1972: The vapor pressure isotope effect in aqueous systems I. H_2O-D_2O (-64 to 100 deg) and H_2O^{16}-H_2O^{18}(-17 to 16 deg); ice and liquid II. Alkali metal chloride solution in H_2O and D_2O (-5 to 100 deg). J. Phys. Chem., v. 76, n. 5, p. 743-762.

Rasmussen, D. H., MacKenzie, A. P., Angell, C. A., and Tucker, J. C., 1973: Anomalous heat capacities of supercooled water and heavy water. Science, v. 181, p. 342-344.

Rasmussen, D. H., and MacKenzie, A. P., 1973: Clustering in supercooled water. J. Chem. Physics, v. 59, n. 9, p. 5003-5013.

Salomon, M., 1969: Thermodynamic properties of liquid H_2O and D_2O and their mixtures. NASA TND-5223 (NTIS), 12 p.

Scherer, J. R., Go, M. K., and Kint, S., 1974: Raman spectra and structure of water from -10 to 90°C. J. Phys. Chem., v. 79, n. 13, p. 1304-1313.

Sofer, Z., and Gat, J. R., 1972: Activities and concentrations of oxygen-18 in concentrated aqueous salt solutions; analytical and geophysical implications. Earth Planet. Sci. Lett., v. 15, n. 3, p. 232-238.

Truesdell, A. H., 1974: In aqueous salt solutions at elevated temperatures--consequences for isotope geochemistry. Earth Planet. Sci. Lett., v. 23, n. 3, p. 387-396.

Wenzel, J., Linderstrom-Lang, C. U., and Rice, S. A., 1975: Amorphous solid water--a neutron diffraction study. Science, v. 187, n. 4175, p. 428-430.

Van Hook, A., 1971: Isotope effect on the thermodynamic activity of water. Knoxville, Tennessee Water Resources Research Center, Rpt. n. 20, 88 p.

7. WATER, WATER VAPOR AND ICE

Bottinga, Y., and Craig, H., 1968: High temperature liquid vapor fractionation factors for H_2O-HDO-H_2O^{18}. Am. Geophys. Union, Trans., v. 49, n. 1, p. 356-357.

Coplen, T. B., and Hanshaw, B. B., 1971: Isotopic fractionation of water by ultrafiltration (abs.). EOS, Am. Geophys. Union, Trans., v. 52, n. 4, p. 364.

Majoube, M., 1970: Fractionation factor of O^{18} between water vapour and ice. Nature, v. 226, n. 5252, p. 1242.

Stewart, M. K., and Friedman, I., 1973: Deuterium fractionation between aqueous salt solutions and water vapor (abs.). EOS, Am. Geophys. Union, Trans., v. 54, n. 4, p. 486.

8. CARBON DIOXIDE AND CARBON DIOXIDE-WATER

Bigeleisen, J., and Harris, T. H., 1973: Fundamental studies in isotope chemistry. Rochester Univ., Rpt. COO-3498-9, 10 p.

Bottinga, Y., and Craig, H., 1969: Oxygen isotope fractionation between CO_2 and water and the isotopic composition of marine atmospheric CO_2. Earth Planet. Sci. Lett., v. 5, n. 5, p. 285-295.

Grootes, P. M., Mook, W. G., and Vogel, J. C., 1968: Isotopic fractionation between gaseous and condensed carbon dioxide. Z. Physik., v. 211, p. 257-273.

Gunter, B. D., and Gleason, J. D., 1971: Isotopic fractionation during gas chromatographic separations. J. Chromatog. Sci., v. 9, n. 3, p. 191-192.

Hoering, T. C., 1961: The effect of physical changes on isotopic fractionation. Carnegie Inst. Washington Year Book 60, p. 201-204.

9. BIOLOGY AND ECOLOGY

Atomic Energy of Canada, Ltd., 1970: Biology and health physics division progress report, April 1 to June 30, 1970. Atomic Energy Canada, Ltd., Chalk River Nuclear Lab., Rpt. PR-B-86, 52 p.

Cortecci, G., and Longinelli, A., 1971: O^{18}/O^{16} ratios in sulfate from living marine organisms. Earth Planet. Sci. Lett., v. 11, n. 4, p. 273-276.

Fujihara, M. P., and O'Brien, R. T., 1965: Toxicity of deuterium oxide to the freshwater fish *Aequidans portalagrensis*--survival of embryos and young fish. U.S. Atomic Energy Comm., HW-SA-3627, 10 p.

Gleason, J. D., and Friedman, I., 1970: Deuterium--natural variations used as a biological tracer. Science, v. 169, p. 1085-1086.

Hotopp, W., and Foerstel, H., 1973: Acclimatized wind track for measurement of oxygen isotope fractionation in isolated leaves and in leaf models. Kernforschungsanlage Juelich G.M.B.H. (F.R. Germany), Inst. Fuer Physikalische Chemie (NTIS), 45 p.

Kritchevsky, D., 1965: Influence of deuterium oxide on biological systems. A. Virus growth in tissue culture cells; B. oxidation of cholesterol by rat liver mitochondria. Philadelphia, Pa., Wistar Inst. Anatomy Bio., Progress Rpt. 1 (NTIS), 19 p.

Kutyurin, V. M., 1969: The isotopic composition of oxygen liberated by plants (in Problems of Geochemistry). Jerusalem, Israel, Program Sci. Transl., p. 661-667.

Mann, L. R. B., 1969: Chemical effects of D_2O on *Escherichia Coli*. Univ. California, Doctoral Thesis (NTIS), 123 p.

Neher, G. H., and Siegel, B. V., 1970: Ultrastructure of ^{19}S antibody-producing cells from als, Pha, and D_2O treated animals. Univ. Oregon, Dept. Pathology (NTIS), 13 p.

Powers, E. L., 1971: Hydroxyl radical and the hydrated electron in radiation effects in cells. Univ. Texas, Austin, Lab. of Radiation Bio. (NTIS), 22 p.

Siegel, S. M., 1971: The performance and capabilities of terrestrial organisms in extreme and unusual gaseous and liquid environments. Univ. Hawaii, Botanical Sci. Paper n. 21 (NTIS), 25 p.

Siegel, S. M., 1973: Toxicology--mechanisms of deuterium oxide action. Univ. Hawaii, Botany Dept., Paper n. 35, (NTIS), 52 p.

Smith, B. N., 1972: Natural abundance of the stable isotopes of carbon in biological systems. Bio. Sci., v. 22, n. 4, p. 226-231.

Stansell, M. J., Mojica, L., and Malvin, H. H., 1967: Measurement of trace levels of deuterium oxide in biologic fluids using infrared spectrophotometry. Brooks AFB, Texas, School of Aerospace Medicine (NTIS), 20 p.

Tulis, J. J., and Figelsbach, H. T., 1970: Isolation and properties of a deuterium-tolerant mutant of *Pasteurella tularensis*. Infection and Immunity, v. 1, n. 1, p. 56-60.

Wershaw, R. L., Friedman, I., Heller, S. J., and Frank, P. A., 1970: Hydrogen isotopic fractionation of water passing through trees. 3rd International Conference on Organic Geochemistry, London, England, September 1966, Proc., p. 55-67.

10. EXTRATERRESTRIAL STUDIES

A. MOON

Boato, G., 1953: The primeval cosmic abundance of deuterium (*in* National Research Council Comm. Nuclear Sci., Proc., Conference of Nuclear Processes in Geologic Settings). September 1953, Chapter 12, p. 45-47.

Bruner, E. C., Jr., and Wilson, T. E., 1969: Deuterium in the earth's exosphere. J. Geophys. Res., v. 74, n. 26, p. 6491-6493.

Clayton, R. N., Onuma, N., and Mayeda, T., 1971: Oxygen isotope fractionation in Apollo 12 rocks and soils (*in* Lunar Sci. Conf., 2nd proc., v. 2). Geochim. Cosmochim. Acta, Suppl. n. 2, p. 1417-1420.

Clayton, R. N., Hurd, J. M., and Mayeda, T. K., 1972: Oxygen isotope abundance in Apollo 14 and 15 rocks and minerals, abs. (*in* Lunar Science III). Lunar Sci. Inst. Contrib. n. 88, p. 141-143.

Clayton, R. N., Hurd, J. M., and Mayeda, T. K., 1972: Oxygen isotopic compositions and oxygen concentrations of Apollo 14 and Apollo 15 rocks and soil (*in* Lunar Science Conference, 3rd Proceedings). Geochim. Cosmochim. Acta, Suppl. n. 3, v. 2, p. 1455-1463.

Clayton, R. N., 1972: Oxygen isotope composition of the Luna 16 soil (*in* The Study of Some Luna 16 Samples). Earth Planet. Sci. Lett., v. 16, n. 2, p. 455-456.

Clayton, R. N., Hurd, J. M., and Mayeda, T. K., 1973: Oxygen isotopic compositions of Apollo 15 and 16 rocks, abs. (in Lunar Science IX). Houston, Texas, Lunar Sci. Inst., p. 148-150.

Clayton, R. N., 1973: Oxygen isotopic composition of the Luna 20 soil (Luna 20--a study of samples from the lunar highlands returned by the unmanned Luna 20 spacecraft). Geochim. Cosmochim. Acta, v. 37, n. 4, p. 811-813.

Clayton, R. N., Grossman, L., and Mayeda, T. K., 1973: A component of primitive nuclear composition in carbonaceous meteorites. Sci. AAAS, v. 182, n. 4111, p. 485-488.

Clayton, R. N., Mayeda, T. K., and Hurd, J. M., 1973: Oxygen isotopic fractionation within ultrabasic clasts of lunar breccia 15445. J. Geol., v. 81, n. 2, p. 227-228.

Clayton, R. N., Mayeda, T. K., and Hurd, J. M., 1974: Loss of O, Si, S and K from the lunar regolith (in Lunar Science V, abs). Houston, Texas, Lunar Sci. Inst., p. 129-131.

Epstein, S., and Taylor, H. P., Jr., 1970: O^{18}/O^{16}, Si^{30}/Si^{28}, D/H and C^{13}/C^{12} studies of lunar rocks and minerals. Science, AAAS, v. 167, n. 3918, p. 533-535.

Epstein, S., and Taylor, H. P., Jr., 1970: The concentration and isotopic composition of hydrogen, carbon and silicon in Apollo 11 lunar rocks and minerals (in Apollo 11 Lunar Sci. Conf., 1970, Houston, Texas, Proc., v. 2, Chemical and Isotope Analysis). Geochim. Cosmochim. Acta Suppl. 1, p. 1085-1096.

Epstein, S., and Taylor, H. P., Jr., 1971: O^{18}/O^{16}, Si^{30}/Si^{28}, D/H and C^{13}/C^{12} ratios in lunar samples (in Lunar Science Conf., 2nd Proc., v. 2). Geochim. Cosmochim. Acta, Suppl. n. 2, p. 1421-1441.

Epstein, S., and Taylor, H. P., Jr., 1971: D/H, O^{18}/O^{16}, C^{13}/C^{12}, Si^{30}/Si^{28} studies in Apollo 11 and 12, lunar rocks and minerals (in Lunar Sci. Conf., abs). NASA, p. 267-268.

Epstein, S., and Taylor, H. P., Jr., 1972: O^{18}/O^{16}, Si^{30}/Si^{28}, C^{13}/C^{12}, and D/H studies of Apollo 14 and 15 moon samples (in Lunar Science Conf., 3rd Proc.). Geochim. Cosmochim. Acta, Suppl. n. 3, v. 2, p. 1429-1454.

Epstein, S., and Taylor, H. P., Jr., 1972: O^{18}/O^{16}, Si^{30}/Si^{28}, C^{13}/C^{12}, and D/H studies of Apollo 14 and 15 samples (in Lunar Science III). Lunar Sci. Inst. Contrib. n. 88, p. 236-238.

Epstein, S., and Taylor, H. P., Jr., 1973: O^{18}/O^{16}, Si^{30}/Si^{28}, C^{13}/C^{12}, D/H and hydrogen and carbon concentration data on Apollo 17 soil (in Apollo 17 Results). EOS, Am. Geophys. Union, Trans., v. 54, n. 6, p. 585-586.

Epstein, S., and Taylor, H. P., Jr., 1973: O^{18}/O^{16}, Si^{30}/Si^{28}, C^{13}/C^{12}, and D/H studies on Apollo 16 lunar samples (in Lunar Science IV, Abstracts). Houston, Texas, Lunar Sci. Inst., p. 228-230.

Epstein, S., and Taylor, H. P., Jr., 1974: Oxygen, silicon, carbon and hydrogen isotope fractionation processes in lunar surface materials, abs. (in Lunar Science V, Abstracts). Houston, Texas, Lunar Sci. Inst., p. 212-214.

Friedman, I., Gleason, J. D., and Hardcastle, K. G., 1970: Water, hydrogen, deuterium, carbon and C^{13} content of selected lunar material (in Apollo 11 Lunar Science Conference, 1970, Houston, Texas, Proceedings, v. 2--Chemical and Isotope Analysis). Geochim. Cosmochim. Acta, Suppl. 1, p. 1103-1109.

Friedman, I., O'Neil, J. R., Adami, L. H., et al., 1970: Water, hydrogen, deuterium, carbon, carbon-13, and oxygen-18 content of selected lunar material. Science, AAAS, v. 167, n. 3918, p. 538-540.

Friedman, I., O'Neil, J. R., Gleason, J. D., et al., 1971: The carbon and hydrogen content and isotopic composition of some Apollo 12 materials (in Lunar Science Conference, 2nd Proceedings, v. 2). Geochim. Cosmochim. Acta, Suppl. n. 2, p. 1407-1415.

Friedman, I., Hardcastle, K. G., and Gleason, J. D., 1972: Isotopic composition of carbon and hydrogen in some Apollo 14 and 15 samples (in The Apollo 15 Lunar Samples). Houston, Texas, Lunar Sci. Inst., p. 302-306.

Friedman, I., Hardcastle, K. G., and Gleason, J. D., 1974: Water and carbon in rusty lunar rock 66095. Science, AAAS, v. 185, n. 4185, p. 346-349.

Friedman, I., Hardcastle, K. G., and Gleason, J. D., 1974: Isotopic composition of carbon and hydrogen in some Apollo 14 and 15 lunar samples. J. Res. USGS, v. 2, n. 1, p. 7-12.

Grossman, L., Clayton, R. N., and Mayeda, T. K., 1974: Oxygen isotopic compositions of lunar soils and Allende inclusions and the origin of the moon, abs. (in Lunar Science V, Abstracts). Houston, Texas, Lunar Sci. Inst., p. 298-300.

Javoy, M., and Fourcade, S., 1973: O^{18}/O^{16} ratios in lunar rocks and fines, abs. (in Europaeisches Kolloquium fuer Geochronologie, III, referate). Fortschr. Mineral., v. 50, Beih. 3, p. 84-85.

Javoy, M., Marette, J., and Pineau, F., 1973: O^{18}/O^{16} ratios in Luna 16 fines. Geochim. Cosmochim. Acta, v. 37, n. 9, p. 2017-2019.

Javoy, M., and Fourcade, S., 1974: O^{18}/O^{16} ratios in lunar fines and rocks. Earth Planet. Sci. Lett., v. 21, n. 4, p. 377-382.

Merlivat, L., Nief, G., and Roth, E., 1972: Deuterium content of lunar materials (in Lunar Science Conference, 3rd Proceedings). Geochim. Cosmochim. Acta, Suppl. n. 3, v. 2, p. 1473-1477.

Merlivat, L., Nief, G., and Roth, E., 1972: Deuterium analysis of hydrogen extracted from lunar material, abs. (in Lunar Science III). Houston, Texas, Lunar Sci. Inst., Contrib. n. 88, p. 537-539.

O'Neil, J. R., and Adami, L. H., 1970: Oxygen isotope analysis of selected Apollo 11 materials (in Apollo 11 Lunar Science Conference, 1970, Houston, Texas, Proc. v. 2--Chemical and Isotope Analysis). Geochim. Cosmochim. Acta,=Suppl. 1, p. 1425-1427.

Onuma, N., Clayton, R. N., and Mayeda, T. K., 1970: Oxygen isotope fractionation between minerals and an estimate of the temperature of formation. Science, AAAS, v. 167, n. 3918, p. 536-538.

Onuma, N., Clayton, R. N., and Mayeda, T. K., 1971: Oxygen isotope fractionation in Apollo 12 rocks and soils, abs. (in Lunar Science Conference 1971). NASA, p. 269.

Taylor, H. P., Jr., and Epstein, S., 1970: O^{18}/O^{16} ratios of Apollo 11 lunar rocks of minerals (in Apollo 11 Lunar Science Conference, 1970, Houston, Texas, Proceedings, v. 2--Chemical and Isotope Analysis). Geochim. Cosmochim. Acta, Suppl. 1, p. 1613-1626.

Taylor, H. P., Jr., and Epstein, S., 1970: Oxygen and silicon isotope ratios of lunar rock 12013. Earth Planet. Sci. Lett., v. 9, n. 2, p. 208-210.

Taylor, H. P., Jr., and Epstein, S., 1973: Oxygen and silicon isotope ratios of the Luna 20 soil (in Luna 20, A Study of Samples from the Lunar Highlands Returned by the Unmanned Luna 20 Spacecraft). Geochim. Cosmochim. Acta, v. 37, n. 4, p. 1107-1109.

B. METEORITES

Boato, G., 1954: The isotopic composition of hydrogen and carbon in the carbonaceous chondrites. Geochim. Cosmochim. Acta, v. 6, n. 5-6, p. 209-220.

Edwards, G., 1953: Hydrogen and deuterium in iron meteorites, sum. Chapter 13 (in National Research Council Comm. Nuclear Science, Proceedings of the Conference on Nuclear Processes in Geologic Settings), p. 48-49.

Fisher, D. E., 1963: The fluorine content of some chondritic meteorites. J. Geophys. Res., v. 68, n. 23, p. 6331-6335.

Friedman, I., 1955: Deuterium content of water in tektites, abs. Geol. Soc. Am. Bull., v. 66, n. 12, pt. 2, p. 1562.

Friedman, I., 1956: Water in tektites (in Nuclear Processes in Geologic Settings). National Academy Science, National Res. Council Pub. 400, p. 1-6.

Deuterium

Garlick, G. D., Naeser, C. W., and O'Neil, J. R., 1971: A Cuban tektite. Geochim. Cosmochim. Acta, v. 35, n. 7, p. 731-734.

Oxygen

Heinzinger, K., Junge, C., and Schidlowski, M., 1971: Oxygen isotope ratios in the crust of iron meteorites. Z. Naturforsch, v. 26a, n. 9, p. 1485-1490.

Lavrukhina, A. K., 1965: Effekty yadernykh reaktsiy v kamennykh meteoritakh (Effects of nuclear reactions in stone meteorites), (in Problemy Geokhimii-Vinogradov Jubilee Volume). Moscow, Izdatel'stvo "Nauka," p. 9-19.

Morgan, J. W., and Ehmann, W. D., 1970: Precise determination of oxygen and silicon in chondritic meteorites by 14-MeV neutron activation with a single transfer system (French and German summaries). Anal. Chim. Acta, v. 49, n. 2, p. 287-299.

Onuma, N., Clayton, R. N., and Mayeda, T. K., 1970: Oxygen isotope fractionation between coexisting materials in chondritic meteorites, abs. EOS, Am. Geophys. Union, Trans., v. 51, n. 4, p. 340.

Onuma, N., Clayton, R. N., and Mayeda, T. K., 1972: Oxygen isotope cosmothermometer. Geochim. Cosmochim. Acta, v. 36, n. 2, p. 169-188.

Onuma, N., Clayton, R. N., and Mayeda, T. K., 1972: Oxygen isotope temperatures of "equilibrated" ordinary chondrites. Geochim. Cosmochim. Acta, v. 36, n. 2, p. 157-168.

Reuter, J. H., Epstein, S., and Taylor, H. P., 1964: O^{18}/O^{16} ratios of some chondritic meteorite and terrestrial ultramafic rocks, abs. Am. Geophys. Union, Trans., v. 45, n. 1, p. 112-113.

Reuter, J. H., Epstein, S., and Taylor, H. P., Jr., 1965: O^{18}/O^{16} ratios of some chondritic meteorites and terrestrial ultramafic rocks. Geochim. Cosmochim. Acta, v. 29, n. 5, p. 481-488.

Reynold, J. H., 1967: Isotopic abundance anomalies in the solar system (in Annual Review of Nuclear Science, v. 17). Palo Alto, California, Annual Reviews, Inc., p. 253-316.

Taylor, H. P., Jr., and Epstein, S., 1964: Comparison of oxygen isotope analyses of tektites, soils and impactite glasses (in Isotopic and Cosmic Chemistry). Amsterdam, North Holland Publishing Co., Chap. 14, p. 181-199.

Taylor, H. P., Jr., Duke, M. B., Silver, L. T., and Epstein, S., 1965: Oxygen isotope studies of minerals in stony meteorites. Geochim. Cosmochim. Acta, v. 29, n. 5, p. 489-512.

Taylor, H. P., Jr., and Epstein, S., 1966: Oxygen isotope studies of Ivory Coast tektites and impactite glass from the Bosumtwi Crater, Ghana. Science, AAAS, v. 153, n. 3732, p. 173-175.

Taylor, H. P., Jr., 1967: Stable isotope studies of ultramafic rocks and meteorites (in Ultramafic and Related Rocks). New York, London, John Wiley and Sons, p. 362-372.

Taylor, H. P., Jr., and Epstein, S., 1962: Oxygen isotope studies on the origin of tektites. J. Geophys. Res., v. 67, n. 11, p. 4485-4490.

Taylor, H. P., Jr., and Epstein, S., 1963: Comparison of O^{18}/O^{16} ratios in tektites, soils and impactite glasses, abs. Am. Geophys. Union, Trans., v. 44, n. 1, p. 93.

Taylor, H. P., Jr., Duke, M. B., Silver, L. T., and Epstein, S., 1964: Oxygen isotopic studies of minerals in stony meteorites, abs. Am. Geophys. Union, Trans., v. 45, n. 1, p. 112.

Taylor, H. P., Jr., and Epstein, S., 1969: Correlations between O^{18}/O^{16} ratios and chemical compositions of tektites. J. Geophys. Res., v. 74, n. 27, p. 6834-6844.

Taylor, H. P., Jr., and Epstein, S., 1973: Correlations between O^{18}/O^{16} ratios and chemical compositions of tektites (in Tektites). Stroudsburg, Dowden, Hutchinson and Ross, p. 181-191.

Vinogradov, A. P., Dontsova, E. I., and Chupakhin, M. S., 1958: The isotopic composition of oxygen in igneous rocks and meteorites. Geochemistry (Geokhimiya), n. 3, p. 235-239.

Vinogradov, A. P., 1958: Meteorites and the earth's crust (Geochemistry of Isotopes) (in U.N. Internat. Conf. on Peaceful Uses of Atomic Energy, 2nd, Geneva). V. 2, p. 255-269.

Oxygen, sulfur and carbon

Vinogradov, A. P., Dontsova, E. I., and Chupakhin, M. S., 1960: Isotopic ratios of oxygen in meteorites and igneous rocks. Geochim. Cosmochim. Acta, v. 18, n. 3/4, p. 278-293.

Walter, L. S., and Clayton, R. N., 1967: Oxygen isotopes-experimental vapor fractionation and variations in tektites. Science, v. 156, n. 3780, p. 1357-1358.

C. SOLAR SYSTEM AND UNIVERSE

Beer, R., Farmer, C. B., Norton, R. N., et al., 1972: Jupiter--observations of deuterated methane in the atmosphere. Science, AAAS, v. 175, n. 4028, p. 1360-1361.

Hayakawa, S., 1960: On the origin of deuterium in the solar system. Tokyo, Astronomical Soc. of Japan, Publ. v. 12, n. 1, p. 115-116.

Holzer, T. E., and Axford, W. I., 1970: Solar wind ion composition. J. Geophys. Res., v. 75, n. 31, p. 6354-6359.

Kuzhevskii, B. M., 1966: O generatsii iader deiteriia v kosmicheskikh luchakh (Generation of deuterium nuclei in cosmic rays). Moscow, Ladernaia Fizika, v. 4, n. 1, p. 130-131.

McElroy, M. B., and Hunten, D. M., 1969: The ratio of deuterium to hydrogen in the Venus atmosphere. J. Geophys. Res., v. 74, n. 7, p. 1720-1739.

Onuma, N., Clayton, R. N., and Mayeda, T. K., 1971: Oxygen isotope effects in the early solar nebula, abs. EOS, Am. Geophys. Union, Trans., v. 52, n. 4, p. 270.

Pasachoff, J. M., and Fowler, W. A., 1974: Deuterium in the universe. Sci. Am., v. 230, n. 5, p. 108-118.

Ramaty, R., and Lingenfelter, R. E., 1969: Cosmic ray deuterium and helium-3 of secondary origin and the residual modulation of cosmic rays. Astrophysical Jour., v. 155, n. 2, pt. 1, p. 587-608.

Schatzman, E., 1967: Cosmogony of the solar system and origin of the deuterium. Paris, Annales d'Astrophysique, v. 30, n. 6, p. 963-973.

Shirk, J. S., Haseltine, W. A., and Pimentel, G. C., 1965: Sinton bands--evidence for deuterated water on Mars. Science, v. 147, n. 3653, p. 48-49.

Smith-Rose, R. L., 1962: Protection of the deuterium line frequency for radio astronomy. Internat. Sci. Radio Union, Bull. n. 134, p. 34-38.

Weinrab, S., 1962: A new upper limit to the galactic deuterium-to-hydrogen ratio. Nature, v. 195, n. 4839, p. 367-368.

11. CONCENTRATIONS IN ROCKS AND MINERALS

A. GENERAL

Bottinga, Y., 1968: Isotope fractionation in the system--calcite-graphite-carbon dioxide-methane-hydrogen-water, abs. Dissert. Abs., Sec. B, Sci. Eng., v. 29, n. 2, p. 662B-663B.

Bottinga, Y., 1969: Calculated fractionation for carbon and hydrogen isotope exchange in the system calcite-carbon dioxide-graphite-methane-hydrogen-water vapor. Geochim. Cosmochim. Acta, v. 33, n. 1, p. 49-64.

Clayton, R. N., and Epstein, S., 1958: The relationship between O^{18}/O^{16} ratios in coexisting quartz, carbonate, and iron oxides from various geological deposits. J. Geol., v. 66, n. 4, p. 352-373.

Clayton, R. N., and Epstein, S., 1961: The use of oxygen isotopes in high-temperature geological thermometry. J. Geol., v. 69, n. 4, p. 447-452.

Clayton, R. N., 1962: Equilibrium constants for oxygen-isotope exchange in mineral systems, abs. J. Geophys. Res., v. 67, n. 4, p. 1632.

Clayton, R. N., 1963: High-temperature isotopic thermometry. U.S. National Acad. Sci., National Res. Coun. Pub. 1075, Nuclear Ser. Rpt. n. 38, p. 185-195.

Clayton, R. N., 1966: Oxygen isotope exchange in rock-water systems, abs. Am. Geophys. Union, Trans., v. 47, n. 1, p. 203.

Dontsova, Ye. I., 1968: Distribution of oxygen isotopes in minerals and rocks and the problem of their genesis. Int. Geol. Congr., 23rd, Czech., Rep., Sect. 6, Proc., p. 67-78.

Dontsova, Ye. I., 1970: Izotopnyy domen kisloroda v protsesse obrazovaniya pored (Oxygen isotope exchange in the process of rock formation), English Summary. Geokhimiya, n. 8, p. 903-916.

Epstein, S., and Taylor, H. P., Jr., 1967: Variation of O^{18}/O^{16} in minerals and rocks (<u>in</u> Researches in Geochemistry, v. 2). New York, John Wiley and Sons, p. 29-62.

Fourcade, S., and Javoy, M., 1973: La geochimie isotopique de l'oxygène dans la région de l'In Ouzzal (Hoggar) (Oxygen isotope geochemistry in the In Ouzzal region, Ahaggar, Sahara), abs. Réun. Annu. Sci. Terre Programme Résumés, p. 92.

Homma, K., and Sakai, H., 1968: Study of oxygen isotope ratios in rocks, I, abs. Geol. Soc. Japan, J., v. 74, n. 2, p. 90.

Zscherpe, G., 1965: Theoretische betrachtungen über den unterschiedlichen einfluss der festkörpendiffusion von isotopen auf ihr mischungsverhältnis in mineralien und gesteinen (Theoretical considerations on the different influence of solid state diffusion of isotopes on their proportions in minerals and rocks). Geophysik u. Geologie, n. 7, p. 57-64.

B. MINERALS

General

Boström, K., 1970: Deposition of manganese-rich sediments during glacial periods. Nature, v. 226, n. 5246, p. 629-630.

Brandt, S. B., 1974: K termodinamike izotopnago domena v ravnovesnykh mineral'nykh assotsiatsiyakh (Thermodynamics of isotope exchange in mineral associations at equilibrium). Akad. Nauk SSSR, Dokl., v. 214, n. 6, p. 1419-1422.

Friedrichsen, H., 1971: Oxygen isotope fractionation between coexisting minerals of the Grimstad-granite. Neues Jahrb. Mineral., Monatsh., n. 1, p. 26-33.

Giletti, B. J., and Anderson, T. F., 1972: Diffusion of oxygen in phlogopite, abs. Geol. Soc. Am., Abs., v. 4, n. 7, p. 517.

Godfrey, J. D., 1962: The deuterium content of hydrous minerals from the east-central Sierra Nevada and Yosemite National Park. Geochim. Cosmochim. Acta, v. 26, p. 1215-1245.

Henderson, J. H., 1971: I. Quartz and cristobalite origin in selected soils and sediments and in relation to montmorillonite origin in bentonites determined by oxygen isotope abundance; II. Cation and silica relationship to vermicalite formation from mica in calcareous harps soil. Univ. Wisconsin, Doctoral.

Hoefs, J., 1969: Natural calcium oxalate with heavy carbon. Nature, v. 223, n. 5204, p. 396.

Kuroda, Y., Suzuoki, T., Matsuo, S., et al., 1974: D/H fractionation of coexisting biotite and hornblende in some granitic rock masses (Japanese summary). Jap. Assoc. Mineral., Petrol., Econ. Geol. J., v. 69, n. 3, p. 95-102.

Lopez de Azcoma, J. M., 1953: Interpretación geofísica de las variaciónes isotopicas naturales de los elementos químicos (Geophysical interpretation of natural isotopic variations of chemical elements). Rev. Cienc. Apl., ano 7, fasc. 3, n. 32, p. 193-202.

Matysh, I. V., Litovchenko, A. S., Kalinichenko, A. M., et al., 1972: Spektry YAMR F^{19}i H^{1} iskusstvennykh ftorflogopitakh (NMR spectra of F^{19} and H^{1} in synthetic fluorphlogopite), English summary. Konst. Svoystva Miner., n. 6, p. 74-76.

Savin, S. M., 1970: Oxygen isotope studies of minerals in ocean sediments. Sci. Council Japan and International Symposium Hydrogeochemistry, Tokyo, Abstracts of Papers, p. 37.

Savin, S. M., 1973: Oxygen and hydrogen isotope studies of minerals in ocean sediments (in Symposium on Hydrogeochemistry, v. 1). Washington, D.C., Clarke Co., p. 372-391.

Svec, H. J., 1962: The absolute isotopic abundance of Cr, V, Li and O in some minerals. Iowa State Univ., Sci. and Tech., Ames, Inst. for Atomic Res., Rpt. Is-500, p. C-39-C-43.

Taylor, H. P., and Epstein, S., 1960: O^{18}/O^{16} ratios in coexisting minerals of igneous rocks, abs. J. Geophys. Res., v. 65, n. 8, p. 2528.

Quartz

Churchman, G. J., Clayton, R. N., Sridhar, K., et al., 1973: Oxygen isotope abundance in quartz from shales, abs. EOS, Am. Geophys. Union, Trans., v. 54, n. 4, p. 489.

Clayton, R. N., Rex, R. W., Syers, J. K., et al., 1972: Oxygen isotope abundance in quartz from Pacific pelagic sediments. J. Geophys. Res., v. 77, n. 21, p. 3907-3915.

Keith, M. L., et al., 1953: Ratio of oxygen isotopes in quartz of contrasted origin, abs. Geol. Soc. Am. Bull., v. 63, n. 12, pt. 2, p. 1270.

Milovskiy, A. V., and Dontsova, Ye. I., 1968: Izotopnyye kriterii genezisa prirodynkh obrazovaniy SiO_2 (Isotopic criteria for the genesis of natural SiO_2 formations). English summary. Geokhim., n. 8, p. 914-921.

Rex, R. W., Syers, J. K., Jackson, M. L., and Clayton, R. N., 1969: Eolian origin of quartz in soils of Hawaiian Islands and in the Pacific Ocean pelagic sediments. Science, v. 163, n. 3864, p. 277-279.

Syers, J. K., Chapman, S. L., Jackson, M. L., Rex, R. W., and Clayton, R. N., 1968: Quartz isolation from rocks, sediments and soils for the determination of oxygen isotopes composition. Geochim. Cosmochim. Acta, v. 32, n. 9, p. 1022-1025.

Quartz and Other Minerals

Blattner, P., and Bird, G. W., 1974: Oxygen isotope fractionation between quartz and K-feldspar at 600°C. Earth Planet. Sci. Lett., v. 23, n. 1, p. 21-27.

Clayton, R. N., 1955: Oxygen isotope abundances in quartz and calcite, abs. Am. Geophys. Union, Trans., v. 36, n. 3, p. 506.

Perry, E. C., Jr., and Bonnichsen, B., 1966: Quartz and magnetite-oxygen-18-oxygen-16 fractionation in metamorphosed Biwabik iron formation. Science, v. 153, n. 3735, p. 528-529.

Yeh, H.-W., and Savin, S. M., 1974: Quartz-clay isotopic gelthermometer for temperature between 180 and 80°C, abs. (in North Central Section, 8th Annual Meeting). Geol. Soc. Am., Abstr., v. 6, n. 6, p. 555.

Chert

Knauth, L. P., and Epstein, S., 1971: Oxygen and hydrogen isotope relationships in cherts and implications regarding the isotopic history of the hydrosphere, abs. Geol. Soc. Am., Abstr., v. 3, n. 7, p. 624.

Knauth, L. P., and Epstein, S., 1973: Hydrogen and oxygen isotope ratios in cherts from the JOIDES deep sea drilling project, abs. Geol. Soc. Am., Abstr., v. 5, n. 7, p. 694-695.

Knauth, L. P., and Epstein, S., 1973: Implications of the isotopic composition of chert for the climatic temperature history of the earth, abs. Geol. Soc. Am., Abstr., v. 5, n. 7, p. 695.

Knauth, L. P., 1973: Oxygen and hydrogen isotope ratios in cherts and related rocks, abs. Diss. Abstr. Int., v. 34, n. 2, p. 718B-719B.

Kolodny, Y., and Epstein, S., 1974: The stable isotope record of DSDP cherts, abs. EOS, Am. Geophys. Union, Trans., v. 55, n. 4, p. 456.

O'Neil, J. R., 1971: O^{18}/O^{16} ratios of cherts associated with the alkaline lakes of east Africa, abs. EOS, Am. Geophys. Union, Trans., v. 52, n. 4, p. 365.

O'Neil, J. R., and Hay, R. L., 1973: O^{18}/O^{16} ratios in cherts associated with the saline lake deposits of east Africa. Earth Planet. Sci. Lett., v. 19, n. 2, p. 257-266.

Perry, E. C., Jr., 1967: The oxygen isotope chemistry of ancient cherts. Earth Planet. Sci. Lett., v. 3, n. 1, p. 62-66.

Perry, E. C., Jr., 1968: Oxygen isotope composition of ancient bedded cherts. Geo. Soc. Am. Spec. Paper 115, p. 175.

Perry, E. C., Jr., and Tan, F. C., 1972: Significance of oxygen and carbon isotope variations in early Precambrian chert and carbonate rocks of southern Africa. Geol. Soc. Am. Bull., v. 83, n. 3, p. 647-664.

Mineral Water Systems

Addy, S. K., and Garlick, G. D., 1974: Oxygen isotope fractionation between rutile and water. Contrib. Mineral. Petrol., Beitr. Mineral. Petrol., v. 45, n. 2, p. 119-121; also Lamont-Doherty Geol. Obs. Contrib. n. 1492, and Mar. Biomed. Inst., Earth Planet. Sci. Div., Contrib. n. 41.

Bertenrath, R., Friedrichsen, H., and Hellner, E., 1973: Die fraktionierung der sauerstoffisotope O^{18}/O^{16} im system eisenoxid-wasser (Fractionation of the oxygen isotopes O^{18}/O^{16} in the system iron oxide-water), abs. (in Deutsche Mineralogische Gesellschaft, Sektion Geochimie, Frueh Jahrstagung, 1972, referate). Fortschr. Mineral, v. 50, Beih. 3, p. 32-33.

Bottinya, Y., and Javoy, M., 1973: Comments on oxygen isotope geothermometry. Earth Planet. Sci. Lett., v. 20, n. 2, p. 250-265.

Clayton, R. N., O'Neil, J. R., and Mayeda, T. K., 1972: Oxygen isotope exchange between quartz and water. J. Geophys. Res., v. 77, n. 17, p. 3057-3067.

Dontsova, Ye. I., and Naumov, G. B., 1967: Opredeleniye temperatur obrazovaniya gidrotermal'nykh kvartsev po izotopnym otnosheniyam kisloroda (Determination of the temperature of formation of hydrothermal quartzes from the oxygen isotopic ratios), English summary. Geokhimiya, n. 5, p. 553-558.

Ingerson, E., and Ueskhau, R. L., 1965: Fraktsionirovaniye izotopov kisloroda v sisteme kvarts-voda (Fractionation of oxygen isotopes in the system quartz-water. Geokhimiya, n. 8, p. 944-960.

O'Neil, J. R., and Clayton, R. N., 1964: Oxygen isotope geothermometry (in Isotopic and Cosmic Chemistry). Amsterdam, North Holland Publishing Co., p. 157-168.

O'Neil, J. R., and Taylor, H. P., Jr., 1965: The oxygen isotope chemistry of feldspar, abs. Am. Geophys. Union, Trans., v. 46, n. 1, p. 170.

O'Neil, J. R., and Taylor, H. P., Jr., 1966: Oxygen isotope equilibrium between muscovite and water, abs. Am. Geophys. Union, Trans., v. 47, n. 1, p. 212-213.

O'Neil, J. R., and Taylor, H. P., Jr., 1967: The oxygen isotope and cation exchange chemistry of feldspars. Am. Mineralogist, v. 52, n. 9-10, p. 1414-1437.

O'Neil, J. R., and Taylor, H. P., Jr., 1969: Oxygen isotope equilibrium between muscovite and water. J. Geophys. Res., v. 74, n. 25, p. 6012-6022.

Suzuoki, T., and Epstein, S., 1970: Hydrogen isotope fractionation factors (α's) between muscovite, biotite, hornblende and water, abs. EOS, Am. Geophys. Union, Trans., v. 51, n. 4, p. 451-452.

Suzuoki, T., and Epstein, S., 1972: Partition of hydrogen isotopes between hydrated silicate minerals and water. Chikyukagaku (Geochem.), v. 5, n. 1, p. 38-44.

Wershaw, R. L., 1964: Oxygen isotope fractionation in the system quartz-water, abs. Dissert. Abs., v. 24, n. 12, pt. 1, p. 5340-5341.

C. SEDIMENTARY ROCKS

General

Cita, M. B., 1972: Studi sul Pliocene e sugli strati di passaggio dal miocene al pliocene; I. Il significato della trasgressione pliocenica alla luce delle nuove scoperte hel Mediterraneo (Studies on the Pliocene and the Miocene/Pliocene transition beds; the significance of Pliocene transgression in the light of the new data from the Mediterranean region). Riv. Ital. Paleontol. Stratigr., v. 78, n. 3, p. 527-596.

Degans, E. T., and Epstein, S., 1962: Relationship between O^{18}/O^{16} ratios in coexisting carbonates, cherts and diatomites. Am. Assoc. Petrol. Geol., Bull., v. 46, n. 4, p. 534-542.

Fontes, J.-C., Fritz, P., Gauthier, J., and Kulbicki, G., 1967: Minéraux argileux éléments--traces et compositions isotopiques (O^{18}/O^{16} et C^{13}/C^{12}) dans les formations gypsifères de l'eocène supérieur et de l'oligocène de Cormeilles-en-Parisis (Clay minerals, trace elements and isotopic compositions of $O^{18}/^{16}$ and $C^{13}/^{12}$ in the gypsiferous formations of the upper Eocene and of the Oligocene of Cormeilles-en-Parisis). Centre Recherches Pau Bull., v. 1, n. 2, p. 315-366.

Fontes, J.-C., Létolle, R., and Nesteroff, W. D., 1973: Les forages DSDP en Méditerranée (leg 13); reconnaissance isotopiques (Deep sea drilling program traverses, leg 13, in the Mediterranean Sea; reconnaissance of isotopes). (in The Mediterranean Sea, A Natural Sedimentation Laboratory), English summary. Stroudsburg, Pennsylvania, Dowden, Hutchinson and Ross, p. 671-680.

Garrels, R. M., and MacKenzie, F. T., 1974: Chemical history of the oceans deduced from post-depositional changes in sedimentary rocks (in Studies in Paleo-oceanography). Based on symposium sponsored by the Society of Economic Paleontologists and Mineralogists. Soc. Econ. Paleontol. Mineral., Spec. Publ. n. 20, p. 193-204.

Herman, Y., 1972: South Pacific Quaternary paleo-oceanography, abs. (in Marine Geology and Geophysics, Géologie et Géophysique du Milieu Marine, Section 8, Int. Géol. Congr., Proc.). Congr. Géol. Int., Programme n. 24, p. 123.

Levandowski, D. W., Kaley, M. E., Silverman, S. R., et al., 1973: Cementation in Lyons sandstone and its role in oil accumulation, Denver Basin, Colorado. Am. Assoc. Pet. Geol. Bull., v. 57, n. 11, pt. 1, p. 2217-2244.

Lloyd, R. M., and Hsu, K. J., 1972: Stable-isotope investigations of sediments from the DSDP III cruise to south Atlantic. Sedimentology, v. 19, n. 1-2, p. 45-58.

Lloyd, R. M., and Hsu, J. K., 1972: Preliminary isotopic investigation of samples from deep sea drilling cruise to the Mediterranean (in The Mediterranean Sea: A Natural Sedimentation Laboratory), French summary. Stroudsburg, Pennsylvania, Dowden, Hutchinson and Ross, p. 681-686.

Savin, S. M., and Epstein, S., 1968: Oxygen and hydrogen isotope variations in sedimentary rocks and minerals, abs. Geol. Soc. Am., Spec. Paper 101, p. 190.

Savin, S. M., 1968: Oxygen and hydrogen isotope ratios in sedimentary rocks and minerals, abs. Dissert. Abs., Sec. B, Sci. and Eng., v. 28, n. 7, p. 2906B.

Savin, S. M., and Epstein, S., 1968: Oxygen and hydrogen isotope geochemistry of ocean sediments. Geol. Soc. Am. Spec. Paper 115, p. 194.

Savin, S. M., and Epstein, S., 1970: The oxygen and hydrogen isotope geochemistry of ocean sediments and shales. Geochim. Cosmochim. Acta, v. 34, n. 1, p. 43-63.

Savin, S. M., and Epstein, S., 1970: The oxygen isotopic compositions of coarse grained sedimentary rocks and minerals. Geochim. Cosmochim. Acta, v. 34, n. 3, p. 323-329.

Savin, S. M., 1974: Isotopic studies of deep sea sediments. AAPG SEPM Annu. Abs., v. 1, p. 77-78.

van Donk, J., and Broecker, W. S., 1969: O^{18}/O^{16} record for the past 450,000 years in the Carribbean deep sea core V12-122, abs. Am. Geophys. Union, Trans., v. 50, n. 4, p. 350.

van Donk, J., 1970: The oxygen isotope record in deep sea sediments, abs. Am. Quaternary Assoc., 1st Meeting, Seattle, Washington, p. 136.

van Donk, J., 1973: A complete Pleistocene O^{18} record for the Atlantic Ocean, abs. Geol. Soc. Am., Abs., v. 5, n. 7, p. 847.

Clay Minerals and Hydroxide Soils

Coplen, T. B., and Hanshaw, B. B., 1973: Ultrafiltration by a compacted clay membrane; oxygen and hydrogen isotope fractionation. Geochim. Cosmochim. Acta, v. 37, n. 10, p. 2295-2310.

Jakubowski, M., and Sapula, J., 1969: Application of isotopic methods for determining physical properties of soils in the structure of the geotechnical cross section, English trans. Nukleonika, v. 13, p. 58-67.

Khitrov, L. M., and Zadorozhnyy, V. I., 1961: Fractionation of isotope oxygen in the soil, English trans. Pochvovedeniye, n. 1, p. 5-14.

Lawrence, J. R., 1970: O^{18}/O^{16} and D/H ratios of soils, weathering zones and clay deposits, abs. Diss. Abstr. Int., v. 31, n. 6, p. 3482-3483B.

Lawrence, J. R., and Taylor, H. P., Jr., 1971: Deuterium and oxygen-18 correlation; clay minerals and hydroxides in Quaternary soils compared to meteoric waters. Geochim. Cosmochim. Acta, v. 35, n. 10, p. 993-1003.

Montana, Idaho, Hawaii, Gulf Coast

Lawrence, J. R., and Taylor, H. P., Jr., 1972: Hydrogen and oxygen isotope systematics in weathering profiles. Geochim. Cosmochim. Acta, v. 36, n. 12, p. 1377-1393.

Russell, J. D., Farmer, V. C., and Velde, B., 1970: Replacement of OH by OD in layer silicates, and identification of the vibrations of these groups in infra-red spectra. Mineral. Mag., v. 37, n. 292, p. 867-879.

Savin, S. M., and Epstein, S., 1970: The oxygen and hydrogen isotope geochemistry of clay minerals. Geochim. Cosmochim. Acta, v. 34, n. 1, p. 25-42.

Stewart, G. L., 1972: Clay-water interaction, the behavior of H^3 and H^2 in adsorbed water, and the isotope effect. Soil Sci. Am. Proc., v. 36, n. 3, p. 421-426.

Syers, J. K., Jackson, M. L., Berkheiser, V. E., Clayton, R. N., and Rex, R. W., 1969: Eolian sediment influence on pedogenesis during the Quaternary. Soil Sci., v. 107, n. 6, p. 421-427.

Yen, H.-W., and Savin, S. M., 1973: The mechanism of burial diagenetic reactions in argillaceous sediments; part 3, oxygen isotopic evidence, abs. EOS, Am. Geophys. Union, Trans., v. 54, n. 4, p. 508.

Evaporites

General

Borishevskiy, Yu. A., and Kristianov, V. K., 1965: Izotopnyy sostav kristallizatsionnoy vody solyanykh mineralov (Isotopic composition of the water of crystallization of saline minerals), English summary. Geokhimiya, n. 7, p. 844-850.

Butler, G. P., Krouse, R. H., and Mitchell, R., 1972: Isotope geochemistry of modern arid supratidal (Sabkha) evaporite environment, Abu Dhabi, Trucial Coast. Am. Assoc. Petrol. Geol. Bull., v. 56, n. 3, p. 607.

Fontes, J.-C., 1966: Interêt en géologie d'une étude isotopique de l'évaporation; cas de l'eau de mer (Geologic interest of an isotopic study of evaporation; case of sea water). Acad. Sci., C.R., Sér. D, v. 263, n. 25, p. 1950-1953.

Fontes, J.-C., and Nielsen, H., 1966: Isotopes de l'oxygène et du soufre dans le gypse parisien. Acad. Sci., C.R., Sér. D, v. 262, n. 26, p. 2685-2687.

Fontes, J. C., Lepurier, C., Mélieres, F., et al.: Le grande episode évaporitique mediterranean au miocene supérieur; prémices géochimiques minéralogiques et isotopiques d'une étude (The great Upper Miocene evaporite episode in the Mediterranean; geochemical, mineralogical and isotopic premises of a study), abs. Réun. Annu. Sci. Terre, Programme Résumés, p. 188.

Matsuo, S., Friedman, I., and Smith, G. S., 1972: Studies of Quaternary saline lakes; 1. Hydrogen isotope fractionation in saline minerals. Geochim. Cosmochim. Acta, v. 36, n. 4, p. 427-435.

Searles Lake, California

Smith, G. I., Friedman, I., and Matsuo, S., 1970: Salt crystallization temperature in Searles Lake, California (in 50th Anniversary Symposia, Mineralogy and Geochemistry of Nonmarine Evaporites). Mineral. Soc. Am., Spec. Paper n. 3, p. 257-259.

Sulfates

Claypool, G. E., Holser, W. T., Kaplan, I. R., et al., 1972: Sulfur and oxygen isotope geochemistry of evaporite sulfates, abs. Geol. Soc. Am. Abs., v. 4, n. 7, p. 473.

Cortecci, G., and Longinelli, A., 1972: Oxygen-isotope variations in a barite slab from the sea bottom off southern California. Chem. Geol., v. 9, n. 2, p. 113-117.

Cortecci, G., 1973: Analisi isotopica di una formazione evaporitica del Miocene superiore (Isotopic analysis of an upper Miocene evaporite formation). Soc. Ital. Mineral. Petrol. Rend., v. 29, n. 1, p. 3-18.

Fontes, J.-C., 1966: Fractionnement isotopique dans l'eau de cristallisation du sulfate de calcium (Isotopic fractionation in the water of crystallization of calcium sulfate), German, English and Russian summaries. Geol. Rundsch., v. 55, n. 1, p. 172-178.

Fontes, J.-C., and Gonfiantini, R., 1967: Fractionnement isotopique de l'hydrogène dans l'eau de cristallisation du gypse (Isotopic fractionation of hydrogen in the water of crystallization of gypsum). Acad. Sci. Comptes Rendus, Sér. D, v. 265, n. 1, p. 4-6.

Fontes, J.-C., and Gohau, G., 1969: Deux siècles de querelle autour de l'origine du gypse parisien (Two centuries of debate about the origin of the Paris gypsum). G. Sci. Progr., Nature (Paris), n. 3413, p. 336-340.

Lloyd, R. M., 1967: Oxygen-18 composition of oceanic sulfate. Science, v. 156, n. 3779, p. 1228-1231.

Lloyd, R. M., 1967: Oxygen isotope fractionation in the sulfate-water system and δO^{18} of oceanic sulfate, abs. Am. Geophys. Union, Trans., v. 48, n. 1, p. 235-236.

Lloyd, R. M., 1968: Oxygen isotope behavior in the sulfate water system. J. Geophys. Res., v. 73, n. 18, p. 6099-6110.

Longinelli, A., and Craig, H., 1967: Oxygen-18 variations in sulfate ions in sea water and saline lakes. Science, v. 156, n. 3771, p. 56-59.

Longinelli, A., and Cortecci, G., 1970: Isotopic abundance of oxygen and sulfate ions from river water. Earth Planet. Sci. Lett., v. 17, n. 4, p. 376-380.

Longinelli, A., and Cortecci, G., 1970: Composizione isotopica dell'ossigeno nei solfati; techniche di misura (Oxygen isotopic composition of sulfates; method of measurement), English summary. Soc. Ital. Mineral. Petrol. Rend., v. 26, n. 2, p. 733-743.

Longinelli, A., and Cortecci, G., 1972: Composizione isotopica dell'ossigeno e dello zolfo nel solfato disciolto in acque sorgive (Isotopic composition of oxygen and sulfur in dissolvate sulfate in spring water), English summary. Soc. Geol. Ital. Boll., v. 91, n. 4, p. 693-701.

Marowsky, G., 1969: Schwefel- und sauerstoffisotopenanalysen von sulfaten des kupferschiefers (Sulfur and oxygen isotope studies of sulfates from the Permian kupferschiefer), English summary. Contrib. Mineral. Petrology-Beitr. Mineral. Petrologie, v. 23, n. 4, p. 361-366.

Mizutani, Y., and Rafter, T. A., 1969: Oxygen isotopic composition of sulphates; part 4, bacterial fractionation of oxygen isotopes in the reduction of sulphate and in the oxidation of sulphur. N.Z. Jour. Sci., v. 12, n. 1, p. 60-68.

New Zealand

Mizutani, Y., and Rafter, T. A., 1973: Isotopic behavior of sulphate oxygen in the bacterial reduction of sulphate. J. Geochem., v. 6, n. 4, p. 183-191.

Rafter, T. A., and Mizutani, Y., 1967: Oxygen isotope composition of sulphates; part 2, preliminary results on oxygen isotopic variation in sulphates and relationship to their environment and their $\delta^{34}S$ values. N.Z. Jour. Sci., v. 10, n. 3, p. 816-840.

Rafter, T. A., 1967: Oxygen isotopic composition of sulphates; part 1, A method for the extraction of oxygen and its quantitative conversion to carbon dioxide for isotope radiation measurements. N.Z. Jour. Sci., v. 10, n. 2, p. 493-510.

Rösler, H. J., Pilot, J., Harzer, D., and Kruger, P., 1968: Isotopengeochemische untersuchungen (O,S,C) on salinar und sapropelsedimentem mitteleuropas (Isotope geochemical investigations (O,S,C) on evaporite and sapropel sediments from central Europe) (in Internat. Geol. Congr., 23rd 1968, Prague, Proc., Sec. 6, Geochemistry). Prague, Academia, p. 89.

East Germany

Sakai, H., and Matsubaya, O., 1970: Sulfur and oxygen ratios of gypsum and barite in the black ore deposits, abs. (in Internat. Mineralogical Assoc., General Meeting, 7th, Internat. Assoc. on the Genesis of Ore Deposits, 1970, Tokyo-Kyoto, Collected Abs.). Tokyo, Internat. Mineral. Assoc., Internat. Assoc. Genesis Ore Deposits, p. 130.

Sakai, H., Osasi, S., and Tsukagishi, M., 1970: Sulfur and oxygen isotopic geochemistry of sulfate in the black ore deposits of Japan. J. Geochem., Japan, v. 4, n. 1, p. 27-39.

Sakai, H., 1971: Sulfur and oxygen isotopic study of barite concretions from banks of the Japan Sea off the northeast Honshu, Japan. J. Geochem., Geochem. Soc. Japan, v. 5, n. 2, p. 79-93.

Sakai, H., and Matsubaya, O., 1971: Sulfur and oxygen isotopic ratios of gypsum and barite in the black ore deposits of Japan (in Geochemistry and Crystallography of Sulphite Minerals in Hydrothermal Deposits). Internat. Mineral. Assoc., Internat. Assoc. Genesis Ore Deposits, Joint Symposium, Soc. Min. Geol. Japan, Spec. Issue n. 2, p. 80-83.

Sakai, H., 1972: Oxygen isotope ratios of some evaporites from Precambrian to Recent ages. Earth Planet. Sci. Lett. v. 15, n. 2, p. 201-205.

Stewart, M. K., 1974: Hydrogen and oxygen isotope fractionation during crystallization of mirabilite and ice. Geochim. Cosmochim. Acta, v. 38, n. 1, p. 167-172.

Teys, R. V., 1956: Izotopnyy sostav kislorada prirodnykh sul'fatov (Isotopic composition of oxygen of natural sulfates). Geokhimiya, n. 3, p. 28-32.

Phosphates

Kolodny, Y., and Kaplan, I. R., 1970: Carbon and oxygen isotopes in apatite CO and co-existing calcite from sedimentary phosphorite. J. Sediment. Petrology, v. 40, n. 3, p. 954-959.

Longinelli, A., 1965: Oxygen isotopic composition of orthophosphate from shells of living marine organisms. Nature, v. 207, n. 4998, p. 716-719.

Longinelli, A., and Sordi, M., 1966: Oxygen isotope composition of phosphate from shells of some living crustaceans. Nature, v. 211, n. 5050, p. 727-728.

Longinelli, A., and Nuti, S., 1967: Composizione isotopica dell'ossigeno nei fosfato; pt. 1, techica di misura e scala di temperature isotopiche (Oxygen isotope composition in phosphate; pt. 1, technique of measurement and isotope temperature scale), English abstract. Soc. Geol. Italiana Boll., v. 86, n. 2, p. 115-132.

Longinelli, A., 1967: Ratios of oxygen-18:oxygen-16 in phosphate and carbonate from living and fossil marine organisms. Nature, v. 211, n. 5052, p. 923-927.

Longinelli, A., and Nuti, S., 1968: Oxygen isotopic composition of phosphorites from marine formations. Earth Planet. Sci. Lett., v. 5, n. 1, p. 13-16.

Longinelli, A., and Nuti, S., 1968: Oxygen-isotope ratios in phosphate from fossil marine organisms. Science, v. 160, n. 3830, p. 879-882.

Longinelli, A., and Nuti, S., 1968: Composizione isotopica dell'ossigeno nei fosfati; part 2, misure su fossili (Oxygen isotope composition of phosphates; part 2, measurement of fossils), English summary. Soc. Geol. Ital., Boll., v. 87, n. 3, p. 533-547.

Longinelli, A., and Nuti, S., 1969: Composizione isotopica dell'ossigeno di fosforiti di origine marina (Oxygen isotopic composition of marine phosphorites), English abstract. Soc. Geol. Ital., Boll., v. 88, n. 1, p. 19-28.

Longinelli, A., 1971: Problems in isotope palaeoclimatology and micropaleonotology (in The Micropaleontology of the Ocean). Cambridge, University Press, p. 603-609.

Longinelli, A., and Nuti, S., 1972: Revised and extended phosphate-water isotopic temperature scale, abs. EOS, Am. Geophys. Union, Trans., v. 53, n. 4, p. 555.

Longinelli, A., and Nuti, S., 1973: Revised phosphate-water isotopic temperature scale. Earth Planet. Sci. Lett., v. 19, n. 3, p. 373-376.

Longinelli, A., and Nuti, S., 1973: Oxygen isotope measurements of phosphate from fish teeth and bones. Earth Planet. Sci. Lett., v. 20, n. 3, p. 337-340.

Teys, R. V., Gromova, T. S., and Kochetkova, S. N., 1958: Isotopnyy sostav prirodnykh fosfatov (Isotopic composition of natural phosphates). Akad. Nauk SSSR, Dokl., v. 122, n. 6, p. 1057-1060.

Limestone and Dolomite

General

Anderson, T. F., and Schneidermann, N., 1972: Isotope relationships in pelagic limestones from the central Carribbean leg 15, deep sea drilling project, abs. EOS, Am. Geophys. Union, Trans., v. 53, n. 4, p. 555.

Anderson, T. F., 1973: Oxygen and carbon isotope compositions of altered carbonates from the western Pacific, core 53.0, deep sea drilling project. Mar. Geol., v. 15, n. 3, p. 169-180.

Artemov, Yu. M., Knorre, K. G., Strizhov, V. P., and Ustinov, V. I., 1966: Izotopnyye otnosheniya Ca-40/Ca-41 i O-18/O-16 v nekotorykh karbonatnykh porodakh (Ca-40/Ca-41 and O-18/O-16 ratios in some carbonate rocks). Geokhimiya, n. 11, p. 1355-1359.

Badiozamani, K., 1973: The dorag dolomitization model application to the middle Ordivician of Wisconsin. J. Sediment. Petrol., v. 43, n. 4, p. 965-984.

Bausch, W., and Hoefs, J., 1972: Die isotopenzusammensetzung von dolomiten und kalken aus dem sueddeutschen malm (The isotopic composition of upper Jurassic dolomites and limestones from southern Germany), English summary. Contrib. Mineral. Petrol.-Beitr. Mineral. Petrol., v. 37, n. 2, p. 121-130.

Bausch, W., and Hoefs, J., 1973: Die isotopenzusammensetzung von dolomiten und kalken aus dem sueddeutschen malm (Isotope measurements in the dolomites and limestones of the southern German malm), abs. (in Deutsche mineralogische gesellschaft, sektion geochimie, fruehjahrstagung). Referate, Fortschr. Mineral., v. 50, Beih. 3, p. 35.

Behrens, E. W., and Land, Lynton, S., 1972: Subtidal Holocene dolomite, Baffin Bay, Texas. J. Sediment. Petrol., v. 42, n. 1, p. 155-161.

Choquette, P. W., 1968: Marine diagenesis of shallow marine lime-mud sediments--insights from δO-18 and δC-13 data. Science, v. 161, n. 3846, p. 1130-1132.

Illinois, Mississippi River

Clayton, R. N., and Degens, E. T., 1959: Use of carbon isotope analyses of carbonates for differentiating freshwater and marine sediments. Am. Assoc. Petrol. Geol. Bull., v. 43, n. 4, p. 890-897.

Clayton, R. N., Jones, B. F., and Berner, R. A., 1968: Isotope studies of dolomite formation under sedimentary conditions. Geochim. Cosmochim. Acta, v. 32, n. 4, p. 415-432.

Deep Springs Lake, California

Clayton, R. N., Skinner, H. C., Berner, R. A., et al., 1968: Isotopic compositions of Recent south Australian lagoonal carbonates. Geochim. Cosmochim. Acta, v. 32, n. 9, p. 983-988.

Cole, R. D., Picard, M. D., Jensen, M. L., et al., 1973: Stable oxygen isotopic composition of carbonate rocks in the Green River formation, eastern Utah, and western Colorado, abs. Am. Assoc. Petrol. Geol. Bull., v. 57, n. 5, p. 956.

Degens, E. T., and Epstein, S., 1962: Stable isotope studies on marine and continental dolomites from recent and ancient sediments, abs. Geol. Soc. Am., Spec. Paper 68, p. 160-161.

Degens, E. T., 1967: Stable isotope distribution in carbonates (in Developments in Sedimentology, v. 9B Carbonate Rocks, Physical and Chemical Aspects). Amsterdam and New York, Elsevier Publ. Co., p. 193-208.

DeGiovani, W. F., Salat, E., Marini, O. J., et al., 1974: Unusual isotopic composition of carbonates from the Irati formation, Brazil. Geol. Soc. Am. Bull., v. 85, n. 1, p. 41-44.

Parana Basin

Donovan, T. J., Friedman, I., and Gleason, J. D., 1974: Recognition of petroleum-bearing traps by unusual isotopic compositions of carbonate-cemented surface rocks. Geology, v. 2, n. 7, p. 351-354.

Oklahoma

Epstein, S., Graf, D. L., and Degens, E. T., 1964: Oxygen isotope studies on the origin of dolomites (in Isotopic and Cosmic Chemistry). Amsterdam, North Holland Publ. Co., p. 169-180.

Fontes, J.-C., Kulbick, G., and Létolle, R., 1969: Les sondages de l'atoll de Mururoa--apercu géochimique et isotopique de la série carbonatée (Borings of the Mururoa atoll--geochemical and isotopic glance at the carbonate series). Cahiers Pacifique, n. 13, p. 69-74.

Fontes, J.-C., Fritz, P., and Létolle, R., 1970: Composition isotopique, minéralogique et genèse des dolomies du bassin de Paris (Isotopic and mineralogic composition and origin of the dolomites of the Paris basin), English abs. Geochim. Cosmochim. Acta, v. 34, n. 3, p. 279-294.

Fritz, P., 1967: Oxygen and carbon isotopic composition of carbonates from the Jura of southern Germany. Canadian J. Earth Sci., v. 4, n. 6, p. 1247-1267.

Fritz, P., 1971: Geochemical characteristics of dolomites and the O^{18} content of Middle Devonian oceans. Earth Planet. Sci. Lett., v. 11, n. 4, p. 277-282.

Fritz, P., and Jackson, S. A., 1972: Geochemical and isotopic characteristics of Middle Devonian dolomites from Pine Point, northern Canada (in Stratigraphy and Sedimentology, Section 6). Int. Geol. Congr., Proc., n. 24, p. 230-243.

Fritz, P., and Jackson, S. A., 1972: The identification and characterization of dolomites by geochemical and isotopic analyses, abs. Int. Geol. Congr. Abstr., Congr. Geol. Int. Resumes, n. 24, p. 184.

Hladikova, J., Smejkal, V., Smid, B., et al., 1972: The carbon and oxygen isotopes in calcites of the teschenite association and in the Tithonian and Berriasian limestones (Moravskoslezske Beskydy Mts), Czech summary. Czech., Ustred. Ustav. Geol., Vestn., v. 47, n. 6, p. 333-340.

Hladikova, J., Posmourny, K., and Smejkal, V., 1973: Isotopic study of carbon and oxygen of hydrothermal carbonates from Bytiz near Pribram, Czech summary. Cas. Mineral. Geol., v. 18, n. 3, p. 243-251.

Hodgson, W. A., 1966: Carbon and oxygen isotope ratios in diagenetic carbonates from marine sediments. Geochim. Cosmochim. Acta, v. 30, n. 12, p. 1223-1233.

Hoefs, J., 1970: Kohlenstoff-und sauerstoffisotopenzusammensetzung von karbonatkonkretionen verschiedener geologischer alter (Carbon and oxygen isotopic composition of carbonate concretions of various geologic ages), abs. J. Fortschr. Mineral., v. 47, Beih. 1, p. 26.

Kastner, M., and Lawrence, J., 1971: Diagenesis and dolomitization in carbonates from O^{18}/O^{16} ratios of authigenic feldspars, abs. EOS, Am. Geophys. Union, Trans., v. 52, n. 4, p. 365.

Keith, M. L., and Weber, J., 1964: Carbon and oxygen isotopic composition of selected limestones and fossils. Geochim. Cosmochim. Acta, v. 28, n. 11, p. 1787-1816.

Keith, M. L., 1969: Isotopic composition of carbonates from the Karroo and comparison with the Passa Dois group of Brazil, abs. (in Gondwana Stratigraphy IUGS Symposium). Spanish, English. Paris, UNESCO, p. 775-778.

Karroo, South Africa

Kier, J. S., 1973: Primary subtidal dolomierite from Baffin Bay, Texas. Am. Assoc. Petrol. Geol. Bull., v. 57, n. 4, p. 788.

Kolodny, Y., 1973: Stable isotopes in the spurrite-carbonate assemblages of the "Mottled Zone" (Israel) (in Europaeisches Kolloquium fuer Geochronologie II). Referate, Fortschr. Mineral., v. 50, Beih. 3, p. 89-90.

Land, L. S., 1973: Holocene meteoric dolomitization of Pleistocene limestones, north Jamaica, abs. Am. Assoc. Petrol. Geol. Bull., v. 57, n. 4, p. 790.

Lawrence, J. R., 1973: Stable oxygen and carbon isotope variations in bulk carbonates from late Miocene to present, in Tyrrhenian basin-site 132, DSDP leg 13 (in Initial Reports of the Deep Sea Drilling Project). Washington, D.C., U.S. Govt. Printing Office, v. 13, pt. 2, p. 796-798.

Limón, L., Narváez, A., Morales, P., et al., 1971: Determinación de las relaciones isótopicas del carbona y el oxígeno en rocas (Determination of the isotopic relationship of carbon and oxygen in carbonate rocks). Inst. Mex. Petról., Rev., v. 3, n. 1, p. 80-82.

Milliman, J. D., 1966: Submarine lithification of carbonate sediments. Science, v. 153, n. 3739, p. 994-997.

North Atlantic

Murata, K. J., Friedman, I., and Gulbrandsen, R. A., 1972: Geochemistry of carbonate rocks in Phosphoria and related formations of the western phosphate field. U.S. Geol. Surv., Prof. Paper n. 800-D, p. D103-D110.

Idaho, Montana, Utah, Wyoming

Pandey, G. C., Sharma, T., and Misra, R. C., 1969: O^{18}/O^{16} and C^{13}/C^{12} variations in the limestone from calc-zone of Pithoragarh. Curr. Sci., v. 38, n. 21, p. 513-514.

Pandey, G. C., Sharma, T., and Misra, R. C., 1970: O^{18}/O^{16} and C^{13}/C^{12} ratio variations in limestones from the Vindhyan range. Geol. Soc. India, J., v. 11, n. 4, p. 397-399.

Russell, K. L., Deffeyes, K. S., Fowler, G. A., and Lloyd, R. M., 1967: Marine dolomite of unusual isotopic composition. Science, v. 155, n. 3759, p. 189-191.

Oregon coast

Scholle, P., and Kinsman, D. J., 1973: Diagenesis of upper cretaceous chalk from North Sea, England and Northern Ireland, abs. Am. Assoc. Petrol. Geol. Bull., v. 57, n. 4, p. 803-804.

Supko, P., Stoffers, P., and Coplen, T. B., 1973: Authigenic dolomites from Red Sea, abs. Am. Assoc. Petrol. Geol. Bull., v. 57, n. 4, p. 807-808.

Theilig, F., Müller, E. P., and Harzer, D., 1968: Die isotope zusammensetzung des sauerstoffs in karbonatgesteinen der Stassfurt-Serie (perm) und ihre bedeutung für fazies und paläothermometrie (The isotopic composition of oxygen in carbonate rocks of the Stassfurt series (Permian) and its significance for facies and paleothermometry), Russian and English summaries. Z. Angew. Geol., v. 14, n. 10, p. 506-508.

Turi, B., 1970: Carbon and oxygen composition of carbonates in limestone blocks and related geodes from the "Black Pozzolans" formation of the Alban hills. Chem. Geol., v. 5, n. 3, p. 195-205.

Weber, J. N., 1965: Evolution of the oceans and the origin of fine-grained dolomites. Nature, v. 207, n. 5000, p. 930-933.

Weber, J. N., and Keith, M. L., 1963: Isotopic composition and environmental classification of selected limestones and fossils, abs. Geol. Soc. Am. Spec. Paper 73, p. 259-260.

Weber, J. N., 1964: Carbon-oxygen isotopic composition of Flagstaff carbonate rocks and its bearing on the history of Paleocene-Eocene Lake Flagstaff of central Utah. Geochim. Cosmochim. Acta, v. 28, n. 8, p. 1219-1242.

Weber, J. N., Bergenback, R. E., Williams, E. G., and Keith, M. L., 1965: Reconstruction of depositional environments in the Pennsylvanian Vanport basin by carbon isotope ratios. J. Sed. Petrol., v. 35, n. 1, p. 36-48.

Weber, J. N., 1967: Factors affecting the carbon and oxygen isotopic composition of marine carbonate sediments, pt. 1, Bermuda. Am. Jour. Sci., v. 265, n. 7, p. 586-608.

Wedepohl, K. H., 1970: Geochemische daten von sediment-aeren karbonaten und karbonatgesteinen in ihrem faziellen und petrogenetischen aussagewet (Geochemical data on carbonate sediments and carbonate rocks with reference to their facies and petrogenetic interpretation) (in Rezente und fossile karbonatsedimentation, ein symposium), English summary. Geol. Bundesanst., Verh., n. 4, p. 672-705.

Williamson, C. R., and Picard, M. D., 1973: Carbonate petrology of Green River formation, Eocene, Ulinta basin, Utah, abs. Am. Assoc. Petrol. Geol., Bull., v. 57, n. 4, p. 813.

Carbonate Minerals

Degens, E. T., and Epstein, S., 1964: Oxygen and carbon isotope ratios in coexisting calcites and dolomites from recent and ancient sediments. Geochim. Cosmochim. Acta, v. 28, n. 1, p. 23-44.

Epstein, S., Degans, E., and Graf, D. L., 1962: The oxygen isotopic compositions of coexisting calcite and dolomite, abs. J. Geophys. Res., v. 67, n. 4, p. 1636.

Fritz, P., and Smith, D. G. W., 1970: The isotopic composition of secondary dolomites. Geochim. Cosmochim. Acta, v. 34, n. 11, p. 1161-1173.

Carbon and oxygen

Gomberg, D. N., and Bonatti, E., 1970: High-magnesian calcite-leaching of magnesium in the deep sea. Science, v. 168, n. 3938, p. 1451-1453.

Lugovaya, I. P., 1973: Izotopnyy sostav kisloroda karbonatov zheleza i produktov ikh termicheskoy dissotsiatsii (Isotopic composition of oxygen from iron carbonates and from products of their thermal dissociation). Geol. Zh. (Russ. ed.), v. 33, n. 3, p. 42-48.

Magaritz, M., 1973: Precipitation of secondary calcite glacier areas; carbon and oxygen isotopic composition of calcites from Mt. Hermon, Israel, and the European Alps. Earth Planet. Sci. Lett., v. 17, n. 2, p. 385-390.

Masi, U., and Turi, B., 1971: La composizione isotopica dell'ossigeno e del carbonio del carbonato presente nei depositi calcitico-fluoritici Pleistocenici del colli della Farnesina, Rome, e di colle di Pianciano, Bracciano (Isotopic composition of oxygen and carbon in the leistocene calcite and fluorite-bearing deposits of Pianciano, Bracciano and Farnesina hills, Rome), English summary.

Osaki, S., 1973: Carbon and isotopic compositions of Tertiary and Permian dolomites in Japan. Geochem. J., v. 6, n. 4, p. 163-177.

Osaki, S., 1973: Carbon and oxygen isotopic ratios of carbonates filling gaps in the Ogi basalt, Sado-Gashima, Niigata, Japan. Geochem. J., v. 6, n. 4, p. 179-182.

Pandey, G. C., and Sharma, T., 1970: Oxygen and carbon isotope fractionation between dolomite and calcite in lesser Himalayan and Vindhyan limestones. Geochim. Cosmochim. Acta, v. 34, n. 5, p. 625-628.

Panov, B. S., Korchemagin, V. O., and Pilot, I., 1974: Pro izotopniy sklad kisnyu ta vugletsyu karbonativ Donets'kogo baseynu (Isotopic composition of oxygen and carbon from carbonates of the Donets basin), English and Russian summaries. Akad. Nauk Ukr. RSR, Dopov., Ser. B, n. 3, p. 225-227.

Parry, W. T., Reeves, C. C., Jr., and Leach, J. W., 1970: Oxygen and carbon isotopic composition of west Texas lake carbonates. Geochim. Cosmochim. Acta, v. 34, n. 7, p. 825-830.

Perry, E. C., Jr., and Tan, F. C., 1973: Significance of carbon isotope variations in carbonates from the Biwabik iron formation, Minnesota (in Genesis of Precambrian Iron and Manganese Deposits), French summary. UNESCO, Earth Sci. Ser., n. 9, p. 299-305.

Sharma, S. K., and Sharma, T., 1968: O-18/O-16 and C-13/C-12 ratios of some natural carbonates. Geol. Soc. India Bull., v. 5, n. 2, p. 53-55.

Smejkal, V., Hladikova, J., Pisa, M., et al.: Izotopicke slozeni nekterych zilnych kalcitu a dolomitu (Isotope composition of some vein calcite and dolomite), English summary. Cas. Mineral. Geol., v. 16, n. 4, p. 423-434.

Walters, L. J., Jr., Claypool, G. E., and Choquette, P. W., 1972: Reaction rates and O^{18} variation for the carbonate phosphoric acid preparation method. Geochim. Cosmochim. Acta, v. 36, n. 2, p. 129-140.

Igneous and Metamorphic Carbonates

Anderson, T. F., 1973: Oxygen isotope composition of recrystallized carbonates associated with submarine volcanic rocks, abs. Am. Assoc. Petrol. Geol. Bull., v. 57, n. 4, p. 767.

Craig, H., and Craig, V., 1972: Greek marbles, determination of Provenance by isotopic analysis. California Univ. Scripps Inst. Oceanogr. Contrib., v. 42, pt. 2, p. 1125-1127.

Oxygen

Deines, P., and Gold, D. P., 1968: Variability of C^{13} and O^{18} in carbonates from a mica periodotite dike near Dixonville, Pennsylvania, abs. Geol. Soc. Am. Spec. Paper 101, p. 51-52.

Deines, P., 1968: Stable carbon and oxygen isotopes of carbonatite carbonates and their interpretation, abs. Dissert. Abs., Sec. B, Sci. and Eng., v. 29, n. 1, p. 247B.

Deines, P., 1968: The carbon and oxygen isotope composition of carbonates from a mica peridotite dike near Dixonville, Pennsylvania. Geochim. Cosmochim. Acta, v. 32, n. 6, p. 613-625.

Deines, P., and Gold, D. P., 1969: The change in carbon and oxygen isotope composition during contact metamorphism of Trenton limestone by the Mount Royal platon. Geochim. Cosmochim. Acta, v. 33, n. 3, p. 421-424.

Quebec study

Deines, P., and Gold, D. P., 1969: The distribution of carbon and oxygen isotopes in the carbonates of the Oka carbonatite, Quebec, Canada, abs. Canadian Mineralogist, v. 10, pt. 1, p. 131.

Deines, P., 1970: The carbon and oxygen isotope composition of carbonates from the Oka carbonatite complex, Quebec, Canada. Geochim. Cosmochim. Acta, v. 34, n. 11, p. 1199-1225.

Dontsova, Ye. I., Migdisov, A. A., and Ronov, A. B., 1972: K voprosu o prichinakh izmeneniya izotopnogo sostava kisloroda v karbonatnykh tolshchakh osadochnoy obolochki (Causes of the change in the oxygen isotope composition in carbonate rocks of the sedimentary cover), English summary. Geokhim. Akad. Nauk SSSR, n. 11, p. 1317-1324.

Fornaseri, M., and Turi, B., 1969: Carbon and oxygen isotopic composition of carbonates in lavas and ejectites from the Alban Hills, Italy. Contr. Mineralogy Petrology, v. 23, n. 3, p. 244-256.

Fredrichsen, H., 1968: Sauerstoffisotopen einiger minerale der karbonatite des fengebietes, süd Norwegen (Oxygen isotopes of some minerals of the carbonatites of the Fen area, south Norway), English summary. Lithos, v. 1, n. 1, p. 70-75.

Hunt, G. H., and Bolivar, S. L., 1972: Oxygen and carbon isotope ratios of carbonate from kimberlites in Elliott County, Kentucky, abs. (in Southeastern Section, 21st Annual Meeting). Geol. Soc. Am., Abstr., v. 4, n. 2, p. 81-82.

Hunt, G. H., and Engelhardt, R. L., 1973: Carbon and oxygen isotope ratios of carbonate from ultramafic dikes of western Kentucky, abs. (in North Central Section, 7th Annual Meeting). Geol. Soc. Am., Abstr., v. 5, n. 4, p. 324.

O'Neil, J. R., Hedge, C. E., and Jackson, E. D., 1970: Isotopic investigations of xenoliths and host basalts from the Honolulu volcanic series. Earth Planet. Sci. Lett., v. 8, n. 3, p. 253-257.

O'Neil, J. R., and Barnes, I., 1971: C^{13} and O^{18} compositions in some freshwater carbonates associated with ultramafic rocks and serpentinites; western United States. Geochim. Cosmochim. Acta, v. 35, n. 7, p. 687-697.

Schidlowski, M., Stahl, W., and Amstutz, G. C., 1970: Oxygen and carbon isotope abundances in carbonates of spilitic rocks from Glarus, Switzerland. Naturwissenschaften, v. 57, n. 11, p. 542-543.

Schidlowski, M., and Stahl, W., 1971: Kohlenstoff- und sauerstoff-isotopenuntersuchungen an der karbonatfraktion alpiner spilite und serpentinite sowie von weilburgiten des Lahn-Dill-gebietes (Carbon and oxygen isotope studies of the carbonate fraction of spilite and serpentinite from the Alps as well as weinburgite from the Lahn-Dill region), English summary. Neues Jahrb. Mineral., Abh., v. 115, n. 3, p. 252-278.

Sheppard, S. M. F., and Schwarez, H. P., 1966: Isotopic studies and the size of partial equilibrium systems in marbles, abs. Am. Geophys. Union, Trans., v. 47, n. 1, p. 213.

Vermont and Ontario, Canada

Sheppard, S. M. F., and Schwarez, H. P., 1968: C^{13}/C^{12} and O^{18}/O^{16} ratios of coexisting dolomite and calcite from metamorphic rocks, Vermont and Canada, abs. Geol. Soc. Am., Spec. Paper 101, p. 199.

Shieh, Y. N., and Taylor, H. P., Jr., 1969: Oxygen and carbon isotope studies of contact metamorphism of carbonate rocks, abs. EOS, Am. Geophys. Union, Trans., v. 50, n. 4, p. 328.

Shieh, Y. N., and Taylor, H. P., Jr., 1969: Oxygen and carbon isotopes studies of contact metamorphism of carbonate rocks. J. Petrol., v. 10, n. 2, p. 307-331.

Birch Creek, California, and Marble Canyon, Texas

Stahl, W., 1971: Isotopen-analysen an carbonaten und kohlendioxid proben aus dem einflussbereich und der weiteren umgebung des Bramscher intrusivs und an hydrothermalen carbonaten aus dem Siegerland (Isotope analysis of carbonates and carbon dioxide probes from the general area of the Bramsche intrusives and of hydrothermal carbonates from Siegerland) (in Das hoehere oberkarbon von westfalen und das Bramscher Massiv; ein symposium; II das Bramscher Massiv), English and French summaries. Geol. Rheinl. Westfalen, v. 18, p. 430-438.

Togliatt, V., 1968: Studio del contenuto in O-18 e in C-13 di alcune rocce carbonatiche della Liguria orientale (Study of the O-18 and C-13 content of some carbonate rocks of eastern Liguria). Accad. Sci. 1st Bologna Atti, Cl. Sci. Fis. Rend., 1965-66, Ser. D, v. 3, n. 1-2, p. 83-96.

Turi, B., 1969: La composizione isotopica dell'ossigeno e del carbonio dei carbonati presenti nelle vulcaniti de S. Venanzo, Umbria (Oxygen and carbon isotopic composition of carbonates present in the San Venanzo volcanites, Umbria), English summary. Period. Mineral., v. 38, n. 3, p. 589-603.

Travertine

Demovic, R., Hoefs, J., and Wedepohl, K. H., 1972: Geochemische undersuchungen an travertinen des Slowakei (Geochemical investigations of travertines from Slovakia), English summary. Contrib. Mineral. Petrol., Beitr. Mineral. Petrol., v. 37, n. 1, p. 15-28.

Friedman, I., 1966: Isotopic variations during the deposition of travertine at Mammoth Hot Springs, abs. Am. Geophys. Union, Trans., v. 47, n. 1, p. 203-204.

Gonfiantini, R., Panichi, C., and Tongiorgi, E., 1968: Isotopic disequilibrium in travertine deposition. Earth Planet. Sci. Lett., v. 5, n. 1, p. 55-58.

Steven, T. A., and Friedman, I., 1968: The source of travertine in the Creede formation, San Juan Mountains, Colorado (in Geological Survey Research, 1968, Chapter B). U.S. Geol. Survey Prof. Paper 600-B, p. B29-B36.

Carbonate Fractionation

Alexander, T. F., 1969: Self-diffusion of carbon and oxygen in calcite by isotopic exchange with carbon dioxide. J. Geophys. Res., v. 74, n. 15, p. 3918-3932.

Belevtsev, Ya. N., Lugovaya, I. P., and Mel'nik, Yu. P., 1969: Izotopny sostav kislorada rudnykh mineralov zhelezistykh porod Krivogo Roga (Oxygen isotope composition of ore minerals in the Krivoi Rog iron-bearing rocks) (in Problemy Obrazovaniya Zhelezistykh porod Dokembriya). Akad. Nauk USSR, Inst. Geol. Nauk Kiev, p. 271-279.

Ukraine, USSR

Bottinga, Y., 1967: Calculation of fractionation factors for carbon and oxygen isotope exchange in the system calcite-carbon dioxide-water, abs. Am. Geophys. Union, Trans., v. 48, n. 1, p. 236.

Clayton, R. N., 1959: Oxygen isotope fractionation in the system calcium-carbonate-water. J. Chem. Physics, v. 30, n. 5, p. 1246-1250.

Clayton, R. N., 1961: Oxygen isotope fractionation between calcium carbonate and water. J. Chem. Physics, v. 34, n. 3, p. 724-726.

Clayton, R. N., Goldsmith, J. R., Johnson, K. J., et al., 1972: Pressure effect on stable isotope fractionation, abs. EOS, Am. Geophys. Union, Trans., v. 53, n. 4, p. 555.

Cole, R. D., Williamson, C. R., Picard, M. D., et al., 1973: Fractionation of stable oxygen isotopes in carbonate rocks of the Green River formation, eastern Utah and western Colorado, abs. (in Rocky Mountain Section, 26th Annual Meeting). Geol. Soc. Am., Abstr., v. 5, n. 6, p. 472.

Fantidis, J., and Ehhalt, D. H., 1970: Variations of the carbon and oxygen isotopic composition in stalagmites and stalactites; evidence of non-equilibrium isotopic fractionation. Earth Planet. Sci. Lett., v. 10, n. 1, p. 136-144.

Fontes, J.-C., and Lagny, P., 1973: Apercu géochimique et isotopique sur la genèse des facies dolomitiques liés aux surfaces d'émersion du Trias moyen dans la région de Sappada, Belluno, Italie (Geochemistry and isotope studies of the genesis of the dolomitic facies related to the middle Triassic emergent surfaces in the Sappada region, Belluno, Italy, abs. Réun. Annu. Sci. Terre, Programme Résumés, p. 187.

Friedman, I., and Hall, W. E., 1963: Fractionation of O^{18}/O^{16} between coexisting calcite and dolomite. J. Geol., v. 71, n. 2, p. 238-243.

Fritz, P., and Fontes, J.-C., 1966: Fractionnement isotopique pendant l'attaque acide des carbonates naturels. Role de la granulométrie (Isotopic fractionation during acid attack of natural carbonates; role of grain size). Acad. Sci. Comptes Rendus, Ser. D, v. 263, n. 19, p. 1345-1348.

Fritz, P., Folingsbee, E., and Smith, D. G. W., 1969: Oxygen isotope fractionation between protodolomite and water, abs. Geol. Soc. Am., Spec. Paper 121, p. 104-105.

Hamza, M. S., and Broecker, W. S., 1974: Surface effect on the isotopic fractionation between CO_2 and some carbonate minerals. Geochim. Cosmochim. Acta, v. 38, n. 5, p. 669-681.

McCrea, J. M., 1950: On the isotopic chemistry of carbonates and a paleotemperature scale. J. Chem. Phys., v. 18, n. 6, p. 849.

Mook, W. G., 1970: Stabiele isotopen van kohlenstoff und sauerstoff in water en kalk (Stable carbon and oxygen isotopes in water and carbonates). Geol. Mijnbouw, v. 49, n. 5, p. 397-404.

Northrup, D. A., and Clayton, R. N., 1966: Oxygen-isotope fractionations in systems containing dolomite. J. Geol., v. 74, n. 2, p. 174-196.

O'Neil, J. R., and Epstein, S., 1964: Oxygen isotope fractionation between dolomite and calcite, abs. Am. Geophys. Union, Trans., v. 45, n. 1, p. 113.

O'Neil, J. R., and Epstein, S., 1966: Oxygen isotope fractionation in the system dolomite-calcite-carbon dioxide. Science, v. 152, n. 3719, p. 198-201.

O'Neil, J. R., Clayton, R. N., and Mayeda, T. K., 1969: Oxygen isotope fractionation in divalent metal carbonates. J. Chem. Phys., v. 51, n. 12, p. 5547-5558.

Schwarcz, H. P., 1966: Oxygen and carbon isotopic fractionation between coexisting metamorphic calcite and dolomite. J. Geol., v. 74, n. 1, p. 38-48.

Sheppard, S. M. F., and Schwarcz, H. P., 1970: Fractionation of carbon and oxygen isotopes and magnesium between coexisting metamorphic calcite and dolomite. Contrib. Mineral. Petrol., Beitr. Mineral. Petrol., v. 26, n. 3, p. 161-198.

Vermont and Ontario, Canada

Tarutani, T., Clayton, R. N., and Mayeda, T. K., 1969: The effect of polymorphism and magnesium substitution on oxygen isotope fractionation between calcium carbonate and water. Geochim. Cosmochim. Acta, v. 33, n. 8, p. 987-996.

Teis, R. V., 1950: Izotopnyi sostav kisloroda karbonatov i ego temperaturnye izmineniya. Akad. Nauk SSSR, Doklady, T.72, p. 73-76.

Weber, J. N., 1964: Oxygen isotope fractionation between coexisting calcite and dolomite. Science, v. 145, n. 3638, p. 1303-1305.

Utah

Weber, J. N., 1965: Oxygen isotope fractionation between coexisting calcite and dolomite in the freshwater upper carboniferous Freeport formation. Nature, v. 207, n. 5000, p. 972-973.

Utah and Pennsylvania

Wendt, I., 1971: Carbon and oxygen isotope exchange between HCO_3 in saline solution and solid $CaCO_3$. Earth Planet. Sci. Lett., v. 12, n. 4, p. 439-442.

Coral Reefs

Cloud, P. E., Jr., Friedman, I., Sisler, F. D., and Dibeler, V. H., 1958: Microbiological fractionation of the hydrogen isotope, Bahamas. Science, v. 127, n. 3311, p. 1394-1395.

Fontes, J.-C., and Létolle, R., 1973: Approche géochimique et isotopique des conditions de diagenèse et d'épigenèse des carbonates sur un atoll dolomitise, Mururoa (A geochemical and isotopic approach to the conditions of diagenesis and epigenesis of the carbonates on a dolomitized atoll, Mururoa), abs. Réun. Annu. Sci. Terre, Programme Résumés, p. 189.

Gross, M. G., 1962: O^{18}/O^{16} and C^{13}/C^{12} ratios of diagenetically altered limestones in the Bermuda Islands, Bikini and Eniwetok atoll, abs. Geol. Soc. Am., Spec. Paper 68, p. 187.

Gross, M. G., 1964: Variations in the O^{18}/O^{16} and C^{13}/C^{12} ratios of diagenetically altered limestones in the Bermuda Islands. J. Geol., v. 72, n. 2, p. 170-194.

Gross, M. G., 1965: Carbonate deposits on Plantagenet Bank near Bermuda. Geol. Soc. Am. Bull., v. 76, n. 11, p. 1283-1290.

Gross, M. G., and Tracey, J. I., Jr., 1966: Oxygen and carbon isotopic composition of limestones and dolomites, Bikini and Eniwetok atolls. Science, AAAS, v. 151, n. 3714, p. 1082-1084.

Weber, J. N., and Woodhead, P. M., 1969: Factors affecting the carbon and oxygen isotopic composition of marine carbonate sediments; II, Heron Island, Great Barrier Reef, Australia. Geochim. Cosmochim. Acta, v. 33, n. 1, p. 19-38.

Weber, J. N., 1974: Stable isotope studies of shallow-water reef corals collected *in situ* from deeply submerged coral reefs, abs. (in Recent Advances in Carbonate Studies, Abstracts Volume). Fairleigh-Dickinson Univ., West Indies Lab., Spec. Paper 6, p. 42-44.

Weber, J. W., Deines, P., Weber, P. H., and Baker, P., 1975: Stable isotope geochemistry of reef coral $CaCO_3$. EOS, Am. Geophys. Union, Trans., v. 56, n. 6, p. 481.

Brachiopods

Lowenstam, H. A., 1959: O^{18}/O^{16} ratios and Sr and Mg contents of calcareous skeletons of recent fossil brachiopods and their bearing on the history of the oceans (in International Oceanographic Congress, 1st N.Y. preprints of abstracts). Washington, D. C., AAAS, p. 71-73.

Lowenstam, H. A., 1960: O^{18}/O^{16} ratios and Sr and Mg contents in recent and fossil articulate brachiopods and their relationship to the water chemistry, abs. Geol. Soc. Am. Bull., v. 71, n. 12, pt. 2, p. 2065-2066.

Lowenstam, H. A., 1961: Mineralogy, O^{18}/O^{16} ratios and strontium and magnesium contents of recent and fossil brachiopods and their bearing on the history of the oceans. J. Geol., v. 69, n. 3, p. 241-260.

Echinoderm

Weber, J. N., and Raup, D. M., 1966: Fractionation of the stable isotopes of carbon and oxygen in marine calcareous organisms, the Echinoidea, part 1, variation of C^{13} and O^{18} content within individuals. Geochim. Cosmochim. Acta, v. 30, n. 7, p. 681-703.

Weber, J. N., 1968: Fractionation of the stable isotopes of carbon and oxygen in calcareous marine invertebrates, the Asteroidea, Ophiuroidea and Crinoidea. Geochim. Cosmochim. Acta, v. 32, n. 1, p. 33-70.

Weber, J. N., and Raup, D. M., 1968: Comparison of C^{13}/C^{12} and O^{18}/O^{16} in the skeletal calcite of recent and fossil echinoids. J. Paleontology, v. 42, n. 1, p. 37-50.

Foraminifera

Deuser, W. J., 1968: Postdepositional changes in the oxygen isotope ratios of Pleistocene foraminifera tests in the Red Sea. J. Geophys. Res., v. 73, n. 10, p. 3311-3314.

Hecht, A. D., 1972: Morphological variation diversity and stable isotope geochemistry of recent planktonic foraminifera from the North Atlantic, abs. Dissert. Abstr. Internat., v. 32, n. 7, p. 4010B.

Hecht, A. D., and Savin, S. M., 1972: Phenotypic variation and oxygen ratios in recent planktonic foraminifera. J. Foraminiferal Res., v. 2, n. 2, p. 55-67.

Atlantic, Pacific and Indian Oceans

Létolle, R., 1971. Oxygène 18, carbone 13, strontium et magnesium dans une population de *Nummulites laevigatus;* essai d'interprétation et conséquences géochimiques conséquences (Oxygen 18 and carbon 13, strontium and magnesium in a population of *Nummulites laevigatus* interpretation and geochemical consequences), English summary. Rev. Geogr. Phys. Geol. Dyn., v. 13, n. 5, p. 439-446.

Longinelli, A., and Tongiorgi, E., 1964: Oxygen isotope composition of some right-and left-coiled foraminifera. Science, v. 144, n. 3621, p. 1004-1005.

Murata, K. J., Friedman, I., and Madsen, B. M., 1969: Isotopic composition of diagenetic carbonates in marine Miocene formations of California and Oregon. U.S. Geol. Survey Prof. Paper 614-B, p. B1-B24.

Shackleton, N., 1968: Depth of pelagic foraminifera and isotopic changes in Pleistocene oceans. Nature, v. 218, n. 5136, p. 79-80.

Shackleton, N. J., Wiseman, J. D. H., and Buckley, H. A., 1973: Non-equilibrium isotopic fractionation between seawater and planktonic foraminiferal tests. Nature, v. 242, n. 5394, p. 177-179.

Shackleton, N. J., and Opdyke, N. D., 1973: Oxygen isotope and paleomagnetic stratigraphy of equatorial Pacific core V28-238; oxygen isotope temperatures and ice volumes on a 10^5 and 10^6 year scale (in CLIMAP Program, Quat. Res.), Washington Univ., Quat. Res. Cent., v. 3, n. 1, p. 39-55.

Van Donk, J., Saito, T., and Shackleton, N. J., 1973: Oxygen isotopic composition and benthonic and planktonic foraminifera of earliest Pliocene age at site 132-Tyrrhenian Basin, DSDP leg 13 (in Initial Reports of the Deep Sea Drilling Project). Washington, D.C., U.S. Govt. Printing Office, v. 73, pt. 2, p. 798-800.

Vergnaud Grazzini, C., and Létolle, R., 1973: Le rapport O^{18}/O^{16} dans les tests de foraminifères fossiles en Méditérranée (The O^{18}/O^{16} ratio in the tests of fossil foraminifera in the Mediterranean Sea), abs. Réun. Annu. Sci. Terre, Programme Résumés, p. 411.

Vinot-Bertouille, A.-C., and Duplessy, J.-C., 1973: Individual isotope fractionation of carbon and oxygen in benthic foraminifera. Earth Planet. Sci. Lett., v. 18, n. 2, p. 247-252.

Mollusks

Fritz, P., and Poplawski, S., 1974: O^{18} and O^{16} in the shells of freshwater molluscs and their environments. Earth Planet. Sci. Lett., v. 24, n. 1, p. 91-98.

Lloyd, R. M., 1964: Variations in the oxygen and carbon isotope ratios of Florida Bay mollusks and their environmental significance. J. Geol., v. 72, n. 1, p. 84-111.

Weber, J. N., and LeRocque, A., 1963: Isotope ratios in marine mollusk shells after prolonged contact with flowing freshwater. Science, v. 142, n. 3600, p. 1666.

Rivière du Sud, Quebec, Canada

D. IGNEOUS ROCKS

General

Baertschi, P., and Silverman, S. R., 1951: The determination of relative abundances of the oxygen isotopes in silicate rocks. Geochim. Cosmochim. Acta, v. 1, n. 4-6, p. 317-328.

Clayton, R. N., 1972: Oxygen-isotope fractionation in high-temperature rocks; terrestrial and extraterrestrial, abs. Internat. Symp. Hydrogeochem. Biogeochem., Abstr., p. 132.

Dontsova, Ye. I., 1966: O nekotorykh zakonomernostyakh raspredeleniya izotopov kisloroda v izverzhennykh gornykh porodakh (On some regularities of distribution of oxygen in igneous rocks), English summary. Geokhimiy, n. 4, p. 430-442.

Garlick, G. D., 1966: Oxygen isotope fractionation in igneous rocks. Earth Planet. Sci. Lett., v. 1, n. 6, p. 361-368.

Muehlenbachs, K., and Hoering, T. C., 1972: Oxygen isotope geochemistry of some rock from island arcs. Carnegie Inst. Wash., Yearb., n. 71, p. 545-547.

Muehlenbachs, K., 1973: The oxygen isotope geochemistry of acidic rocks from Iceland, abs. EOS, Am. Geophys. Union, Trans., v. 54, n. 4, p. 499-500.

Montigny, R. J. E., 1969: Géochemie isotopique et origine des roches andesitiques étude d'un cas particulier (Isotopic geochemistry and origin of andesite rocks, study of a particular case) (in Symposium on Volcanoes and their Roots, 1969, Oxford). Oxford Univ., Internat. Assoc. Volcanol. Chem. Earth's Interior, Dept. Geol. Mineral., p. 140.

Nesterenko, G. V., and Dontsova, Ye. I., 1967: Fraktionirovaniye izotopov kisloroda v protsesse stanovleniya granitoidov Tyrny-Auza (Fractionation of oxygen isotopes during formation of the Tyrny-Auz granitoids), English summary. Geokhim., n. 12, p. 1445-1452.

Pineau, F., Javoy, M., Craig, H., et al., 1973: O^{18}/O^{16} and C^{13}/C^{12} ratios in volcanic and plutonic rocks from oceanic environment (in Europaeisches Kolloquium fuer Geochronologie, II). Referate, Fortschr. Mineral., v. 50, Beih. 3, p. 119-120.

Schwander, H., 1973: Bestimmung des relativen sauerstoffisotopenverhältnisses in silikatgesteinen und mineralien (Determination of the relative oxygen isotope ratios in silicate rocks and minerals). Geochim. Cosmochim. Acta, v. 4, n. 6, p. 261-291.

Taylor, H. P., Jr., 1968: The oxygen isotope geochemistry of igneous rocks. Contr. Mineralogy and Petrology, v. 19, n. 1, p. 1-71.

Viswanathan, S., 1972: $\delta(O^{18})$ whole rock from $\delta(O^{18})$ quartz and its petrogenetic significance. Curr. Sci., v. 41, n. 22, p. 800-802.

Intrusives

Allegre, C. J., Hamet, J., and Javoy, M., 1968: Etudes des rapports isotopiques Sr^{87}/Sr^{86} et O^{18}/O^{16} du massif granitique du Folat et de ses filons; datation et pétrogenèse (Isotopic relation between Sr^{87}/Sr^{86} and O^{18}/O^{16} in the Folat granitic massif and its veins; dating and petrogenesis). Acad. Sci., Comptes Rendus, Sér. D, v. 266, n. 22, p. 2180-2183.

Anderson, A. T., Jr., 1966: Mineralogy of the Labrieville anorthosite, Quebec. Am. Mineralogist., v. 51, n. 11-12, p. 1671-1711.

Anderson, A. T., Jr., 1966: Oxidation of the LaBlanche Lake titaniferous magnetite deposit, Quebec. J. Geology, v. 76, n. 5, p. 528-547.

Boettcher, A. L., 1967: The Rainy Creek alkaline ultramafic igneous complex near Libby, Montana, pt. 1, Ultramafic rocks and fenite. J. Geol., v. 75, n. 5, p. 526-553.

Conway, C. M., and Taylor, H. P., Jr., 1969: O^{18}/O^{16} and C^{13}/C^{12} ratios of coexisting minerals in the Oka and Maquet Cove carbonatite bodies. J. Geol., v. 77, n. 5, p. 618-626.

Quebec, Canada, and Arkansas

Deines, P., and Gold, D. P., 1973: The isotopic composition of carbonatite and kimberlite carbonates and their bearing on the isotopic composition of deep-seated carbon. Geochim. Cosmochim. Acta, v. 37, n. 7, p. 1709-1733.

Canada and Africa

Denaeyer, M. E., 1970: Rapports isotopiques δO et δC et conditions d'affleurement des carbonatites de l'Afrique centrale (δO and δC isotope ratios and carbonatite outcrops of central Africa). Acad. Sci., Comptes Rendus, Sér. D, v. 270, n. 18, p. 2155-2158.

Dontsova, Ye. I. : Opyt primeneniya izotopnykh sootnosheniy kisloroda v geokhimicheskikh issledovaniyakh (Attempt to use the isotopic ratios of oxygen in geochemical investigations) (in Khimiya zemnoy kory geokhimicheskaga konferentsiya posvyashchennoy stoletiyu so dnya rozhdeniya akademika V.I. Vernadskogo Trudy, V. 2). Akad. Nauk SSSR Inst. Geokhimii i Analit. Khimii, p. 631-637.

Kola Peninsula, USSR

Dontsova, Ye. I., 1967: Application of oxygen isotope ratios in geochemical research. Chem. Earth's Crust, v. 2, Jerusalem, Israel, Program for Scientific Translations, p. 672-679.

Kola Peninsula, USSR

Dontsova, Ye. I., 1968: Sauerstoffisotope und das problem der granitgenese (Oxygen isotopes and the problem of granite genesis). Z. Angew Geol., v. 14, n. 12, p. 633-636.

Dontsova, Ye. I., 1969: O genezise shehelochnykh porod kol'skogo poluostrova v svete dannykh ob izotopnom sostave kisloroda (Genesis of alkaline rocks in the Kola peninsula determined by isotopic oxygen composition), English summary. Geokhim., n. 11, p. 1306-1311.

Dontsova, Ye. I., 1972: Izotopy kisloroda i voprosy genezisa granitov (Oxygen isotopes and the problem of the genesis of granites) (in Ocherki sovremennoy geokhimii i analiticheskoy khimii). Moscow, Izd. Nauka, p. 362-368.

Friedrichsen, H., 1970: Oxygen isotope geothermometry in carbonatites, abs. (in International Mineralogical Assoc., General Meeting, 7th, International Assoc. on the Genesis of Ore Deposits, Tokyo-Kyoto, 1970, collect. abstr.). Internat. Mineral. Assoc., Internat. Assoc. Genesis Ore Deposits, p. 162.

Friedrichsen, H., 1971: Oxygen isotope geothermometry in carbonatites, abs. (in International Mineralogical Assoc., 7th General Meeting, paper and proceedings). Mineral. Soc. Japan, Spec. Paper n. 1, p. 255.

Friedrichsen, H., 1973: Die verteilung der sauerstoffisotope bei der differentiation karbonatischer magmen (Distribution of oxygen isotopes in the differentiation of carbonatite magma) (in Deutche Mineralogische Gesellschaft, Sektion Geochimie, Fruehjahrstagung, 1972). Referate, Fortschr. Mineral., v. 50, Beih. 3, p. 34-35.

Garlick, G. D., MacGregor, I. D., and Vogel, D. E., 1971: Oxygen isotope ratios in eclogites from kimberlites. Science, v. 172, n. 3987, p. 1025-1027.

Hoefs, J., and Epstein, S., 1969: O^{18}/O^{16} ratios of minerals from migmatites, rapakivi granites and orbicular rocks. Lithos, v. 2, n. 1, p. 1-8.

Finland, Mexico, California

Honma, K., and Sakai, H., 1970: Study of oxygen isotopic ratios of rocks; II, study of the Takayama gabbro complex by means of oxygen isotopes, abs. Geol. Soc. Japan, J., v. 76, n. 2, p. 102.

Honma, H., and Sakai, H., 1971: Oxygen isotope study of the Koyama intrusive complex, Yamaguchi-prefecture, southwest Japan. Geochem. J. (Geochem. Soc. Japan), v. 4, n. 3, p. 93-103.

Javoy, M., and Allègre, C. J., 1968: Etude de la composition O^{18}/O^{16} de quelques éclogites; conséquences pétrologiques et géophysiques (Petrologic and geophysical significance of O^{18}/O^{16} composition in certain eclogites). Soc. Géol. Fr., Bull., Sér. 7, v. 9, n. 6, p. 800-808.

Algeria

Javoy, M., 1969: Oxygen isotopic ratios in eclogitic rocks, abs. EOS, Am. Geophys. Union, Trans., v. 50, n. 4, p. 341.

Javoy, M., 1972: O^{18}/O^{16} ratios in ultramafic complexes, abs. EOS, Am. Geophys. Union, Trans., v. 53, n. 4, p. 536.

Javoy, M., Genest, A.-M., and Bottinga, Y., 1973: Fractionnement isotopiques de l'oxygène dans les séries ophiolitiques et les roches spilitiques; implications génétiques (Oxygen isotope fractionation in an ophiolite series and spilites; genetic implications). Réun. Annu. Sci. Terre, Programme Résumés, p. 231.

Matsuhisa, Y., Tainosho, Y., and Matsubaya, O., 1973: Oxygen isotope study of the Ibaragi granitic complex, Osaka, southwest Japan. Geochem. J. (Geochem. Soc. Japan), v. 7, n. 4, p. 201-213.

Pineau, F., and Javoy, M., 1969: Détermination des rapports isotopiques O^{18}/O^{16} et C^{13}/C^{12} dans diverses carbonatites; implications génétiques (Determination of O^{18}/O^{16} and C^{13}/C^{12} isotopic ratios in carbonatites; genetic implications). Acad. Sci., Comptes Rendus, Sér. D, v. 269, n. 20, p. 1930-1933.

Pineau, F., Loubet, M., Javoy, M., et al., 1969: Etude isotopique et géochimique de quelques carbonatites et problème de l'origine de ces roches (Isotopic and geochemical study of some carbonatites and the problem of their origin) (<u>in</u> Symposium on Volcanoes and their Roots, 1969, Oxford). Internat. Assoc. Volcanol. Chem. Earth's Interior, Oxford, Univ., Dept. Geol. Mineral., Vol. Abstr., p. 143.

Pineau, F., Javoy, M., and Allègre, C. J., 1973: Etude systematique des isotopes de l'oxygène, du carbone et du strontium dans les carbonatites (Systematics of the isotopes of oxygen, carbon and strontium in carbonatites), English summary. Geochim. Cosmochim. Acta, v. 37, n. 11, p. 2363-2377.

Canary Island, Sweden, Norway, Morocco, Angola

Pinus, G. V., 1971: Heavy oxygen isotope of the olivine from the ultramafic rocks as indicator of their genesis. J. Geophys. Res., v. 76, n. 5, p. 1339-1341.

Pinus, G. V., and Dontsova, Ye. I., 1971: Izotopnyye otnosheniya kisloroda olivina ul'traosnovnykh porod razlichnogo genezisa (Isotopic ratios of oxygen in olivine from ultramafic rocks of various origins), English summary. Geol. Geofiz., Akad. Nauk SSSR, Sib. Otd., n. 12, p. 3-8.

Schwarcz, H. P., and Clayton, R. N., 1965: Oxygen isotope studies of amphibolites. Canadian J. Earth Sci., v. 2, n. 2, p. 72-84.

United States and Ontario, Canada

Schwarcz, H. P., 1966: Oxygen isotope fractionation between host and exsolved phases in perthite. Geol. Soc. Am., Bull., v. 77, n. 8, p. 879-882.

Shcherbak, N. P., and Lugovaya, I. P., 1972: Ispol'zovaniye izotopov kisloroda dlya resheniya voprosov petrogenezisa granitoidov Ukrainskogo shchita (Application of oxygen isotopes to the solution of problems of petrogenesis of granitoids of the Ukrainian Shield). Geol. Zh. (Russ. ed), v. 32, n. 1, p. 18-27.

Sheppard, S. M. F., and Epstein, S., 1970: D/H and O^{18}/O^{16} ratios of minerals of possible mantle or lower crustal origin. Earth Planet. Sci. Lett., v. 9, n. 3, p. 232-239.

United States, Canada, South Africa, India, Australia

Sheppard, S. M. F., and Dawson, J. B., 1973: C^{13}/C^{12} and D/H isotope variations in "Primary Igneous Carbonatites," abs. (in Europaeisches kolloquium fuer geochronologie, II Referate). Fortschr. Mineral., v. 50, Beih. 3, p. 128-129.

Sims, P. K., and Viswanathan, S., 1972: Giants range batholith (in Geology of Minnesota, A Centennial Volume). St. Paul, Minnesota, Minn. Geol. Surv., p. 120-139.

Spooner, E. T. C., Beckinsale, R. D., Fyfe, W. S., et al., 1974: O^{18} enriched ophiolitic metabasic rocks from E. Liguria, Italy, Pindus, Greece, and Troodos, Cyprus. Contrib. Mineral. Petrol., Beihr. Mineral. Petrol., v. 47, n. 1, p. 41-62.

Suwa, K., Osaki, S., Oana, S., et al., 1969: Isotope geochemistry and petrology of the Mbeya carbonatite, southwestern Tanzania, east Africa (in Nagoya University, African rift valley expedition, 1968). Nagoya Univ., J. Earth Sci., v. 17, p. 125-168.

Taylor, H. P., Jr., Frechen, J., and Degens, E. T., 1967: Oxygen and carbon isotope studies of carbonatites from the Laacher See district, West Germany, and the Alno district, Sweden. Geochim. Cosmochim. Acta, v. 31, n. 3, p. 407-430.

Taylor, H. P., Jr., and Epstein, S., 1960: O^{18}/O^{16} ratios in rocks and coexisting minerals of the Skaergaard intrusion of east Greenland, abs. Geol. Soc. Am., Bull., v. 71, n. 12, pt. 2, p. 1989.

Taylor, H. P., Jr., and Epstein, S., 1963: O^{18}/O^{16} ratios in rocks and coexisting minerals of the Skaergaard intrusion, east Greenland. J. Petrology, v. 4, n. 1, p. 51-74.

Taylor, H. P., Jr., 1969: Oxygen isotope studies of anorthosites, with particular reference to the origin of bodies in the Adirondack Mountains, New York (in Origin of Anorthosite and Related Rocks). New York State Mus. and Sci. Service Mem. 18, p. 111-134.

Taylor, H. P., Jr., and Forester, R. W., 1973: An oxygen and hydrogen isotope study of the Skaergaard Intrusion and its country rocks, abs. EOS, Am. Geophys. Union, Trans., v. 54, n. 4, p. 490.

Turi, B., and Taylor, H. P., Jr., 1971: An oxygen and hydrogen isotope study of a granodiorite pluton from the Southern California batholith. Geochim. Cosmochim. Acta, v. 35, n. 4, p. 383-406.

Turi, B., and Taylor, H. P., Jr., 1971: O^{18}/O^{16} study of the Domenigoni Valley granodiorite pluton, southern California batholith, abs. EOS, Am. Geophys. Union, Trans., v. 52, n. 4, p. 364-365.

Vinogradov, A. P., Dontsova, Ye. I., Gerasimovskiy, V. I., et al., 1971: Izotopnyy sostav kisloroda karbonatitov kontinental'nykh riftovykh zon vostochnoy Afriki (Isotopic composition of oxygen in carbonatites of the continental rift zones of East Africa), English summary. Geokhim. Akad. Nauk SSSR, n. 5, p. 507-514.

Uganda

Viswanathan, S., 1974: Oxygen isotope studies of early Precambrian granitic rocks from the Giants range batholith, northeastern Minnesota. Lithos, v. 7, n. 1, p. 29-34.

Vogel, D. E., and Garlick, G. D., 1969: Oxygen isotope ratios in eclogites, abs. EOS, Am. Geophys. Union, Trans., v. 50, n. 4, p. 342.

Pegmatites

Dontsova, Ye. I., Milovskiy, A. V., and Mel'nikov, F.P.: Fraktsionirovaniye izotopov kisloroda v protsesse obrazovaniya mineralov pegmatitovykh tel (Fractionation of oxygen isotopes during the formation process of minerals in pegmatites), English summary. Geokhim. Akad. Nauk SSSR, n. 5, p. 562-567.

Friedrichsen, H., 1973: O^{18}/O^{16} and H^2/H^1 variations in pegmatites, abs. (<u>in</u> Europaeisches Kolloquium fuer Geochronologie, II Referate). Fortschr. Mineral., v. 50, Beih. 3, p. 70.

Taylor, H. P., Jr., and Epstein, S., 1962: O^{18}/O^{16} ratios of feldspar and quartz in zoned granitic pegmatites, abs. Geol. Soc. Am., Spec. Paper 68, p. 283-284.

Extrusives

Allegre, C. J., and Javoy, M., 1969: Coupled O^{18}/O^{16} and Sr^{87}/Sr^{86} determination in volcanic rocks and magmatologic implications, abs. EOS, Am. Geophys. Union, Trans., v. 50, n. 4, p. 356.

Anderson, A. T., Jr., Clayton, R. N., and Mayeda, T. K., 1971: Oxygen isotope thermometry of mafic igneous rocks. J. Geol., v. 79, n. 6, p. 715-729.

Hawaii, Tahiti, Tristan da Cunha, Ross Island

Cerling, T. E., 1973: Correlation of Plio-Pleistocene tuffs utilizing O^{18}/O^{16} isotope ratios, east Rudolf basin, Kenya. Iowa State Univ., Masters thesis.

Fourcade, S., and Javoy, M., 1971: Anomalous O^{18}/O^{16} ratios in rhyolitic rocks, abs. (in European colloquium of Geochronology). Soc. Geol. Belg., Ann., v. 94, n. 2, p. 113-114.

Friedman, I., and Smith, R. L., 1958: The deuterium content of water in some volcanic glasses. Geochim. Cosmochim. Acta, v. 15, n. 3, p. 218-228.

Olancha, California; Yellowstone National Park; New Mexico; New Zealand

Friedman, I., Smith, R. L., Levin, B., and Moore, A., 1964: The water and deuterium content of phenocysts from rhyolitic lavas (in Isotopic and Cosmic Chemistry). Amsterdam, North Holland Publishing Co., p. 200-204.

Friedman, I., 1967: Water and deuterium in pumice from the 1959-1960 eruption of Kilauea Volcano, Hawaii (in Geological Survey Research 1967, Chap. B). U.S. Geol. Survey, Prof. Paper 575-B, p. B120-B127.

Holmes, A., 1956: The ejectamenta of Katwe crater, southwest Uganda. Koninkl. Netherland. Geol.-Mijnb. Genootschap, Geol. Ser., pt. 16, p. 139-166.

Javoy, M., Hamet, J., and Allegre, C. J., 1969: Sources des roches rhyolitiques; approche isotopique à l'aide des deux traceurs O^{18}/O^{16} et Sr^{87}/Sr^{86} (Sources of rhyolitic rocks; isotopic approach using two tracers, O^{18}/O^{16} and Sr^{87}/Sr^{86}) (in Symposium on Volcanoes and their Roots, 1969, Oxford, vol. abstr.). Internat. Assoc. Volcanol. Chem. Earth's Interior, Oxford Univ., Dept. Geol. Mineral., p. 127-128.

Lambert, S. J., 1975: Stable isotope studies of an active hydrothermal system in New Mexico. EOS, v. 56, n. 6, p. 480.

Lipman, P. W., and Friedman, I., 1974: Oxygen-isotope variations in phenocrysts from compositionally zoned ash-flow sheets, southern Nevada, abs. (in Cordilleran Section, 70th Annual Meeting). Geol. Soc. Am., Abstr. v. 6, n. 3, p. 207.

Lipman, P. W., and Friedman, I., 1975: Interaction of meteoric water with magma; oxygen-isotope study of ash-flow sheets from southern Nevada. Geol. Soc. Am., Bull., v. 56, p. 695-702.

Masi, U., and Turi, B., 1972: Frazionamento isotopico del carbonio e dell'assigeno negli inclusi calcarei metamorfosati del "Tufo grigio compano" acut. di fiano, Salerno (Isotopic fractionation of carbon and oxygen in metamorphosed calcareous inclusions of the "Campan gray tuff" Fiano, Salerno), English summary. Period. Mineral., v. 41, n. 2, p. 291-310.

Matsuhisa, Y., Matsubaya, O., and Sakai, H., 1973: Oxygen isotope variations in magmatic differentiation processes of the volcanic rocks in Japan. Contrib. Mineral. Petrol., Beitr. Mineral. Petrol., v. 39, n. 4, p. 277-288.

Muehlenbachs, K., Anderson, A. T., Jr., and Sigvaldason, G. E., 1972: The origins of O^{18}-poor volcanic rocks from Iceland, abs. EOS, Am. Geophys. Union, Trans., v. 53, n. 4, p. 556.

Noetzlin, J., 1953: La mésure des rapports isotopiques et ses applications possibles à la volcanologie (The measurement of isotope ratios and their possible applications to volcanology). Bull. Volcanolog., Sér. 2, tome 12, p. 115-125.

France and Germany

Oana, S., 1965: Stable isotope studies on volcanic emanations (Japanese with English abstract). Volcanol. Soc. Japan Bull., 10th Anniversary Volume, p. 233-237.

Olive, J. R., and Hermannsson, S., 1967: Proceedings of the Surtsey Research Conference, Reykjavik, June 25-28, 1967. Am. Inst. Biol. Sci., Washington, D.C. (NTIS), 117 p.

Pineau, F., and Javoy, M., 1973: Composition isotopique de l'oxygène dans les roches volcaniques de Tenerife et la complexe plutonique de Fuerteventura, ils Canaries (Oxygen isotope composition of the volcanic rocks of Tenerife and Fuerteventura plutonic complex, Canary Islands, abs.). Réun. Annu. Sci. Terre, Programme Résumes, p. 340.

Stuckless, J. S., and O'Neil, J. R., 1971: Sr^{87}/Sr^{86} and O^{18}/O^{16} ratios in the Superstition-Superior volcanic area, Arizona; implications concerning magma genesis, abs. Geol. Soc. Am., Abstr., v. 3, n. 7, p. 725.

Stuckless, J. S., and O'Neil, J. R., 1973: Petrogenesis of the Superstition-Superior volcanic area as inferred from strontium and oxygen isotope studies. Geol. Soc. Am., Bull., v. 84, n. 6, p. 1987-1997.

Taylor, H. P., Jr., 1967: Origin of red rock granophyres, abs. Am. Geophys. Union, Trans., v. 48, n. 1, p. 245-246.

Wenner, D. B., 1973: O^{18}/O^{16} studies of Cenozoic volcanic rocks in Antarctica, abs. Geol. Soc. Am., Abstr., v. 5, n. 7, p. 859.

E. METAMORPHIC ROCKS

Addy, S. K., and Ypma, P. J. M., 1972: Oxygen isotope and textural criteria suggesting metamorphism of Ducktown ore deposits, Tennessee, abs. EOS, Am. Geophys. Union, Trans., v. 53, n. 11, p. 1128.

Addy, S. K., and Ypma, P. J. M., 1972: Wall rock alternation as the result of regional metamorphic reactions; oxygen isotope and microprobe studies at Ducktown, Tennessee, abs. Geol. Soc. Am., Abs., v. 4, n. 7, p. 430-431.

Anderson, A. T., and Clayton, R. N., 1966: Equilibrium oxygen isotope fractionation between minerals in igneous, meta-igneous and meta-sedimentary rocks, abs. Am. Geophys. Union, Trans., v. 47, n. 1, p. 212.

Anderson, A. T., 1967: The dimensions of oxygen isotope equilibrium attainment during prograde metamorphism. J. Geol., v. 75, n. 3, p. 323-332.

Becker, R., 1973: O^{18}/O^{16} ratios and the temperature of formation of magnetite in the Brockman iron formation, western Australia, abs. (<u>in</u> Deutsche Mineralogische Gesellschaft, Sektion Geochimie, Fruehjahrstagung, 1972, referate). Fortschr. Mineral., v. 50, Beih. 3, p. 33-44.

Belevtsev, Ya. N., Lugovaya, I. P., and Mel'nik, Yu. P., 1967: O vozmozhnosli ispol'zovaniya razlichiy izotopnogo sostava kisloroda dlya ustanovleniya genezisa bogatykh zheleznykh rud Krivogo Rogo (Utilization of differences in isotopic composition of oxygen to determine the origin of the Krivoi Rog iron ores). Akad. Nauk SSSR, Dokl., v. 173, n. 3, p. 678-680.

Ukraine, USSR

Black, P. M., 1974: Oxygen isotope study of metamorphic rocks from the Ouegoa district, New Caledonia. Contrib. Mineral. Petrol., Beitr. Mineral. Petrol., v. 47, n. 3, p. 197-207.

Blattner, P., 1972: Oxygen isotope composition of minerals from Lepontine gneisses, Valle Bodengo, Prov. di Sondrio, Italia. Schweiz. Mineral. Petrogr. Mitt., v. 52, n. 1, p. 33-37.

Blattner, P., and Cooper, A. F., 1974: Carbon and oxygen isotopic composition of carbonatite dikes and metamorphic country rock of the Haast schist terrain, New Zealand. Contrib. Mineral. Petrol., Beitr. Mineral. Petrol., v. 44, n. 1, p. 17-27.

Bottinga, Y., and Javoy, M., 1974: Oxygen isotope geothermometry of igneous and metamorphic rocks, abs. EOS, Am. Geophys. Union, Trans., v. 55, n. 4, p. 477-478.

Brandt, S. B., and Lapidets, I. L., 1970: K. teorii razdeleniya izotopov v kontaktovykh metasomaticheskikh oreolahk (On the theory of isopertition in metasomatic contact halos), English summary. Akad. Nauk SSSR Sib. OTDEL, Inst. Geokhimii Ezhegodnik, p. 200-204.

Brown, P. E., and Miller, J. A., 1969: Interpretation of isotopic ages in orogenic belts (<u>in</u> Time and Place in Orogeny). Geol. Soc. London, Spec. Publ., n. 3, p. 137-155.

Clayton, R. N., 1963: Oxygen isotope geochemistry-thermometry of metamorphic rocks (in Studies in Analytical Geochemistry). Royal Soc. Canada, Spec. Pub. 6, p. 42-57.

Clayton, R. N., 1972: Application of oxygen isotope measurements to problems of iron formations, abs. (in Precambrian Iron Formation Symposium, Abstracts and Field Guides). Duluth, Soc. Econ. Geol., Univ. Minnesota, p. 6.

Cornides, I., Kiss, J., and Szeredai, L., 1966: A középsőmátrai érctelér képződési hőmerséklete az ^{18}O-izotóp relatív gyakorisága alapján (French summary). Földt, Közl., v. 96, n. 1, p. 43-50.

Hungary

Devereux, I., 1968: Oxygen isotope ratios of minerals from the regionally metamorphosed schists of Otago, New Zealand. N. Z. J. Sci., v. 11, n. 3, p. 526-548.

Dontsova, Ye. I., and Milovsky, A. V., 1968: Oxygen isotopes in granitization. Geochem. Internat., v. 4, n. 3, p. 537-544.

Eslinger, E. V., and Savin, S. M., 1971: Oxygen isotope studies of burial metamorphism of the Belt supergroup, Glacier National Park, Montana, abs. Geol. Soc. Am., Abstr., v. 3, n. 7, p. 558-559.

Eslinger, E. V., 1972: Mineralogy and oxygen isotope ratios of hydrothermal and low-grade metamorphic argillaceous rocks, abs. Dissert. Abstr. Inst., v. 32, n. 8, p. 4669B.

Eslinger, E. V., and Savin, S. M., 1973: Oxygen isotope geothermometry of the burial metamorphic rocks of the Precambrian belt supergroup, Glacier National Park, Montana. Geol. Soc. Am., Bull., v. 84, n. 8, p. 2549-2560.

Fourcade, S., 1971: Isotopic temperatures in low pressure granulite facies, abs. (in European Colloquium of Geochronology). Soc. Geol. Belg., Ann., v. 94, n. 2, p. 113.

Fourcade, S., and Javoy, M., 1973: Rapports O^{18}/O^{16} dans les roches du vieux socle catazonal d'In Ouzzal, Sahara Algerien (O^{18}/O^{16} ratios in the rocks of the old catazonal basement of In Ouzzal, Algerian Sahara), English summary. Contrib. Mineral. Petrol., Beitr. Mineral Petrol., v. 42, n. 3, p. 235-244.

Friedrichsen, H., Mueller, G., and Stahl, W., 1973: Sauerstoff isotopenuntersuchungen an mineralen eines metamorphen profils der Hohen Tauern, Oesterreich (English summary). Contrib. Mineral. Petrol., Beitr. Mineral. Petrol., v. 40, n. 2, p. 159-168.

Friedrichsen, H., Mueller, G., and Stahl, W., 1973: Oxygen isotope studies on metamorphosed rocks of the Venediger Massif, Hohe Tauern, abs. (in Europaeisches Kolloquium fuer Geochronologie, II referate). Fortschr. Mineral., v. 50, Beih. 3, p. 70-71.

Garlick, G. D., and Epstein, S., 1964: Oxygen isotope ratios in coexisting metamorphic minerals, abs. Am. Geophys. Union, Trans., v. 45, n. 1, p. 112.

Garlick, G. D., 1965: Oxygen isotope ratios in coexisting minerals of regionally metamorphosed rocks, abs. Dissert. Abstr., v. 25, n. 11, p. 6533.

Garlick, G. D., 1967: Oxygen isotope ratios in coexisting minerals of regionally metamorphosed rocks. Geochim. Cosmochim. Acta, v. 31, n. 2, p. 181-214.

Huang, C.-I., 1975: Isotopic and petrologic studies of contact metasomatic ores at Ely, Nevada. EOS, v. 56, n. 6, p. 480.

James, H. L., and Clayton, R. N., 1962: Oxygen isotope fractionation in metamorphosed iron formations of the Lake Superior region and in other iron-rich rocks (in Petrologic Studies). Geol. Soc. Am., p. 217-239.

Magaritz, M., and Taylor, H. P., Jr., 1974: Oxygen and hydrogen isotope studies of serpentinization in the Troodos ophialite complex, Cyprus. Earth Planet. Sci. Lett., v. 23, n. 1, p. 8-14.

Magaritz, M., and Taylor, H. P., Jr., 1975: O^{18}/O^{16}, D/H and C^{13}/C^{12} studies of the Franciscan formation, California. EOS, v. 56, n. 6, p. 480.

Matsuhisa, Y., Honma, H., Matsubaya, O., et al., 1972: Oxygen isotope study of the Cretaceous granitic rocks in Japan. Contrib. Mineral. Petrol., Beitr. Mineral. Petrol., v. 37, n. 1, p. 65-70.

Mel'nik, Yu. P., 1973: Fiziko-khimicheskiye usloviya obrazovaniya dokembriyskikh zhelezistykh kvartsitov (Physical and chemical conditions of genesis of Precambrian ferruginous quartzites). Akad. Nauk Ukr. SSR, Inst. Geokhim. Fiz. Mineral., 287 p.

Ukraine, USSR

Perry, E. C., 1969: Oxygen isotope relations in progressively metamorphosed rocks adjacent to the Bethlehem granite gneiss, Mascoma quadrangle, New Hampshire, abs. Geol. Soc. Am., Abstr., pt. 7, p. 173.

Perry, E. C., 1970: Carbon and oxygen isotope ratios in 3000 m.y. old rocks of southern Africa, abs. Geol. Soc. Am., Abstr., v. 2, n. 7, p. 649.

Rio, M., Pachiaudi, C., 1973: Apport des mésures de C^{13} et O^{18} à l'étude de la silification (The use of C^{13} and O^{18} measurements in silification studies). Réun. Annu. Sci. Terre, Programme Résumés, p. 361.

Schwarcz, H. P., Clayton, R. N., and Mayeda, T. K., 1962: Oxygen-isotope variations in metamorphosed calcareous rocks of New England, abs. Geol. Soc. Am. Spec. Paper 68, p. 265.

Schwarcz, H. P., Clayton, R. N., and Mayeda, T. K., 1970: Oxygen-isotope studies of calcareous and pelitic metamorphic rocks, New England. Geol. Soc. Am., Bull., v. 81, n. 8, p. 2299-2315.

Sharma, T., Mueller, R. F., and Clayton, R. N., 1965: O^{18}/O^{16} ratios of minerals from the iron formations of Quebec. J. Geol., v. 73, n. 4, p. 664-667.

Sidorenko, A. V., Borshevskiy, Yu. A., Borisova, S. L., et al., 1973: Izotopno-kislorodnaya diagnostika orto-paraamfibolitov (Oxygen isotope diagnosis of orthoamphibolites and paraamphibolites). Akad. Nauk SSSR, Dokl., v. 211, n. 1, p. 208-211.

Kola Peninsula, USSR

Shieh, Y.-N., 1969: Oxygen, carbon and hydrogen isotope studies of contact metamorphism, abs. Dissert. Abstr. Internat., Sec. B, Sci. Eng., v. 30, n. 2, p. 714B.

Shieh, Y.-N., and Taylor, H. P., Jr., 1969: Oxygen and hydrogen isotope studies of contact metamorphism in the Santa Rosa range, Nevada, and other areas. Contr. Mineral. Petrol., v. 20, n. 4, p. 306-356.

Shieh, Y.-N., and Schwarcz, H. P., 1971: Oxygen isotope studies on some granitic rocks from the Grenville province of southeastern Ontario, abs. EOS, Am. Geophys. Union, Trans., v. 52, n. 4, p. 365.

Shieh, Y.-N., Schwarcz, H. P., and Shaw, D. M., 1972: Correlation between O^{18}/O^{16} ratios and chemical compositions of the Apsley gneiss from the Grenville province of Ontario, abs. EOS, Am. Geophys. Union, Trans., v. 53, n. 4, p. 556.

Shieh, Y.-N., and Schwarcz, H. P., 1974: Oxygen isotope studies of granite and migmatite, Grenville province of Ontario, Canada. Geochim. Cosmochim. Acta, v. 38, n. 1, p. 21-45.

Taylor, H. P., Jr., Albee, A. L., and Epstein, S., 1961: O^{18}/O^{16} ratios of coexisting minerals in three mineral assemblages of kyanite-zone pelitic schist, abs. J. Geophys. Res., v. 66, n. 8, p. 2565-2566.

Taylor, H. P., Jr., and Epstein, S., 1962: Relationship between O^{18}/O^{16} ratios in coexisting minerals of igneous and metamorphic rocks, part 2, application to petrologic problems. Geol. Soc. Am., Bull., v. 73, n. 6, p. 675-693.

Taylor, H. P., Jr., and Epstein, S., 1962: Relationship between O^{18}/O^{16} ratios in coexisting minerals of igneous and metamorphic rocks, part 1, principles and experimental results. Geol. Soc. Am., Bull., v. 73, n. 4, p. 361-480.

Southern California and Greenland

Taylor, H. P., Jr., Albee, A. L., and Epstein, S., 1963: O^{18}/O^{16} ratios of coexisting minerals in three assemblages of kyanite-zone pelitic schist. J. Geol., v. 71, n. 4, p. 513-522.

Taylor, H. P., Jr., 1964: Isotopic evidence for large-scale oxygen exchange during metamorphism of Adirondack igneous rocks, abs. Geol. Soc. Am. Spec. Paper 16, p. 163-164.

Taylor, H. P., Jr., 1966: Deuterium-hydrogen ratios in coexisting minerals of metamorphic and igneous rocks, abs. Am. Geophys. Union, Trans., v. 47, n. 1, p. 213.

Taylor, H. P., Jr., and Coleman, R. G., 1968: O^{18}/O^{16} ratios of coexisting minerals in glaucophane-bearing metamorphic rocks. Geol. Soc. Am. Bull., v. 79, n. 12, p. 1727-1755.

California, Oregon and New Caledonia

Turi, B., and Taylor, H. P., Jr., 1971: O^{18}/O^{16} ratios of Johnny Lyon granodiorite and Texas Canyon quartz monzonite plutons, Arizona, and their contact aureoles. Contrib. Mineral. Petrol., Beitr. Mineral. Petrol., v. 32, n. 2, p. 138-146.

Viswanathan, S., Perry, E. C., Jr., and Sums, P. K., 1971: Oxygen isotope studies of early Precambrian granitic and metamorphic rocks from the western part of the Giants range batholith, northeastern Minnesota, abs. Inst. Lake Superior Geol., Abstr. Field Guides, n. 17, p. 66-69.

Vogel, D. E., and Garlick, G. D., 1970: Oxygen isotope ratios in metamorphic eclogites. Contrib. Mineral. Petrol., Beitr. Mineral. Petrol., v. 28, n. 3, p. 183-191.

Wenner, D. B., and Taylor, H. P., Jr., 1969: δD and δO^{18} studies of serpentinization of ultramafic rocks, abs. Geol. Soc. Am. Abstr. w/programs, pt. 7, p. 234-235.

California, Vermont, Pennsylvania, Dominican Republic and Guatemala

Wenner, D. B., and Taylor, H. P., Jr., 1971: Temperatures of serpentinization of ultramafic rocks based on O^{18}/O^{16} fractionation between coexisting serpentine and magnetite. Contrib. Mineral. Petrol., Beitr. Mineral. Petrol., v. 32, n. 3, p. 165-185.

Wenner, D. B., 1971: Hydrogen and oxygen isotopic studies of serpentinization of ultramafic rocks, abs. Dissert. Abstr. Internat., v. 32, n. 3, p. 1667B-1668B.

Wenner, D. B., and Taylor, H. P., Jr., 1972: D/H and O^{18}/O^{16} data on the origin of the H_2O and the temperature involved in serpentinization of ultramafic rocks, abs. Internat. Geol. Congr. Geol. Internat., Resumes, n. 24, p. 58.

Wenner, D. B., and Taylor, H. P., Jr., 1972: O^{18}/O^{16} and D/H studies of a Precambrian granite-rhyolite terrane in S.E. Missouri, abs. EOS, Am. Geophys. Union, Trans., v. 53, n. 4, p. 534.

North America and Caribbean Sea

Wenner, D. B., and Taylor, H. P., Jr., 1973: Oxygen and hydrogen isotope studies of the serpentinization of ultramafic rocks in oceanic environments and continental ophiolite complexes. Am. Jour. Sci., v. 273, n. 3, p. 207-239.

California, Oregon, Atlantic and Pacific Oceans

Wilson, A. F., Green, D. C., and Davidson, L. R., 1970: The use of oxygen isotope geothermometry on the granulites and related intrusives, Musgrave ranges, central Australia. Contrib. Mineral. Petrol., Beitr. Mineral. Petrol., v. 27, n. 2, p. 166-178.

Wilson, A. F., and Green, D. C., 1971: The use of oxygen isotopes for geothermometry of Aoterozoic and Archaean granulites (in Symposium on Archaean Rocks). Geol. Soc. Aust., spec. publ., n. 3, p. 389-400.

F. SECONDARY PROCESSES

Ore-Forming Fluids

Fritz, P., 1969: The oxygen and carbon isotopic composition of carbonates from the Pine Point lead-zinc ore deposits. Econ. Geol., v. 64, n. 7, p. 733-742.

Ohmoto, H., and Rye, R. O., 1968: The ores of the Bluebell mine, British Columbia, a product of meteoric water? Abs. Econ. Geol., v. 63, n. 6, p. 699.

O'Neil, J. R., and Silberman, M. L., 1974: Stable isotope relations in epithermal Au-Ag deposits (in Stable isotopes as applied to problems of ore deposits). Econ. Geol., v. 69, n. 6, p. 902-909.

Roedder, E., 1965: Report on S.E.G. symposium on the chemistry of the ore-forming fluids. Econ. Geol., v. 60, n. 7, p. 1380-1403.

Rye, R. O., 1966: The carbon, hydrogen and oxygen isotope composition of the hydrothermal fluids responsible for the lead-zinc deposits at Aovidencia, Zocatecas, Mexico. Econ. Geol., v. 61, n. 8, p. 1399-1427.

Rye, R. O., Doe, B. R., and Wells, J. D., 1974: Stable isotope and lead isotope study of the Cortez, Nevada, gold deposits and surrounding areas. J. Res. U.S. Geol. Survey, v. 2, n. 1, p. 13-23.

Sugisaki, R., 1968: Origin of the hydrothermal ore fluid and its cooling mechanisms from the viewpoint of oxygen isotope composition, abs. Geol. Soc. Japan, J., v. 74, n. 2, p. 128.

White, D. F., 1974: Diverse origins of hydrothermal ore fluids. Econ. Geol., v. 69, n. 954-973.

Inclusions

Baumann, L., Harzer, D., and Leyeder, O., 1972: Beitrag zum charakter mineralbildender loesungen in cinigen hydrothermallagerstaetten der DDR (The composition of the mineral-forming solutions in the hydrothermal deposits of East Germany). Dtsch. Ges. Geol. Wiss. Ber., Reihe B, Mineral. Lagerstattenforsch., v. 17, n. 3, p. 341-355.

Baumann, L., Harzer, G. D., and Leyeder, O., 1973: Vklad v issledovaniye kharakte mineraloobrazuyushchikh rastvorov na nekotorykh gidrotermalnykh mestorozh deniyakh GDR (Contributions to the study of the nature of mineral-forming solutions in some hydrothermal deposits of East Germany) (in Gidrotermal'nyye Protsessy Mezhdunar). Geokhim. Kongr. Dokl., n. 1, v. 2, p. 364-384.

Hall, W. E., and Friedman, I., 1963: Composition of fluid inclusions, Cave-in-Rock district, Illinois, and upper Mississippi Valley zinc-lead district. Econ. Geol., v. 58, n. 6, p. 886-911.

Hall, W. E., Friedman, I., and Nash, J. T., 1974: Fluid inclusion and light stable isotope study of the Climax molybdenum deposits, Colorado (in Stable Isotopes as Applied to Problems of Ore Deposits). Econ. Geol., v. 69, n. 6, p. 884-901.

Hamilton, L., 1968: Variations in carbon and oxygen isotope ratios as a possible guide to ore. Mineral. Deposita, v. 3, n. 1, p. 81-84.

Kokubu, N., et al., 1961: Deuterium content of minerals, rocks and liquid inclusions from rocks. Geochim. Cosmochim. Acta, v. 21, n. 3/4, p. 247-256.

Landis, G. P., Rye, R. O., and Sawkins, F. J., 1974: Geologic fluid inclusions and stable isotope studies of the Pasto Bueno tungsten-base metal deposits, northern Peru, abs. Econ. Geol., v. 69, n. 7, p. 1025.

Matsui, E., Saliti, E., and Marini, O. J., 1974: D/H and SU O^{18}/O^{16} ratios in waters contained in geodes from the basaltic province of Rio Grande do Sul, Brazil. Geol. Soc. Am., Bull., v. 85, n. 4, p. 577-580.

Ohmoto, H., 1972: Fluid inclusions and istope study of the lead-zinc deposits at the Bluebell mine, British Columbia, Canada, abs. Internat. Symposium, Hydrogeochem. Biogeochem, Abstr., p. 23.

Ohmoto, H., and Rye, R. O., 1974: Hydrogen and oxygen isotopic composition of fluid inclusions in the Kuroko deposits, Japan (in Stable Isotopes as Applied to Problems of Ore Deposits). Econ. Geol., v. 69, n. 6, p. 947-953.

Robinson, B. W., and Ohmoto, H., 1973: Mineralogy, fluid inclusions and stable isotopes of the Echo Bay U-Ni-Ag-Cu deposits, Northwest Territories, Canada. Econ. Geol., v. 68, n. 5, p. 635-656.

Rye, R. O., and O'Neil, J. R., 1968: The O^{18} content of water in primary fluid inclusions from Providencia, North Central Mexico. Econ. Geol., v. 63, n. 3, p. 232-238.

Rye, R. O., and Sawkins, F. J., 1974: Fluid inclusions and stable isotope studies on the Casapalca Ag-Pb-Zn-Cu deposit, central Andes, Peru. Econ. Geol., v. 69, p. 181-205.

Hydrothermal Minerals

Clayton, R. N., Muffler, L. J. P., and White, D. E., 1968: Oxygen isotope study of calcite and silicates of the River Ranch, no. 1 well, Salton Sea geothermal field, California. Am. Jour. Sci., v. 266, n. 10, p. 968-979.

Eslinger, E. V., and Savin, S., 1971: Oxygen isotope studies of hydrothermally altered volcanic rocks from Broadlands, New Zealand, geothermal area, abs. EOS, Am. Geophys. Union, Trans., v. 52, n. 4, p. 365.

Eslinger, E. V., and Savin, S. M., 1971: Oxygen isotope studies of hydrothermal alteration and burial diagenesis, abs. Clay Mineral. Conf., Program Abstr., n. 20, p. 25.

Eslinger, E. V., and Savin, S. M., 1973: Mineralogy and oxygen isotope geochemistry of the hydrothermally altered rocks of the Ohaki-Broadlands, New Zealand, geothermal area. Am. Jour. Sci., v. 273, n. 3, p. 240-267.

Garlick, G. D., and Epstein, S., 1966: The isotopic composition of oxygen and carbon in hydrothermal minerals at Butte, Montana. Econ. Geol., v. 61, n. 8, p. 1325-1335.

Harzer, D., 1967: Sauerstoffisotopenuntersuchungen an varistischen und postvaristischen lagerstätten sachsens (Oxygen isotope investigation of Variscan and post-Variscan mineral deposits in Saxony), with discussion. Freiberg, Forschungsh., Reihe C, n. 209, p. 137-151.

Harzer, D., 1970: Isotopengeochemische untersuchungen ($O^{18}$and C^{13}) an hydrothermalen mineralen, aus ganglagerstaetten der DDR (Geochemical isotope studies, O^{18} and C^{13}, of hydrothermal minerals from vein deposits in East Germany), with Russian and English summaries. Freiberg. Forschungsh., Reihe C, n. 247, 132 p.

Magaritz, M., and Issar, A., 1973: Carbon and oxygen isotopes in epigenetic hydrothermal rocks from Hamam-el-Farun, Sinai. Chem. Geol., v. 12, n. 2, p. 137-146.

Osaki, S., 1973: Carbon and oxygen isotopic compositions of hydrothermal rhodochrosites from Oc. Inakuraishi and Jokaku mines, Hokkaido, Japan. Geochem. J. (Geochem. Soc. Japan), v. 6, n. 4, p. 151-162.

Sugisaki, R., and Jenson, M. L., 1971: Oxygen isotopic studies of silicate minerals with specific reference to hydrothermal mineral deposits. Geochem. J. (Geochem. Soc. Japan), v. 5, n. 1, p. 7-21.

Taylor, H. P., Jr., 1967: Oxygen isotope studies of hydrothermal mineral deposits (in Geochemistry of Hydrothermal Ore Deposits). New York, Holt, Rinehart and Winston, p. 109-142.

Taylor, H. P., Jr., 1973: O^{18}/O^{16} evidence for meteoric hydrothermal alteration and ore deposition in the Tonopah, Comstock Lode, and Goldfield mining districts, Nevada, abs. Geol. Soc. Am., Abstr., v. 5, n. 7, p. 835.

Taylor, H. P., Jr., 1973: The application of oxygen and hydrogen isotope studies to the problem of hydrothermal alteration and ore deposition, abs. Geol. Soc. Am., Abstr., v. 5, n. 7, p. 834-835.

Wodzicki, A., 1972: Mineralogy, geochemistry and origin of hydrothermal alteration and sulphide mineralization in the disseminated molybdenite and skarn-type copper sulphide deposit of Copperstain Creek, Takaka, New Zealand. New Zealand Geol. Geophys., v. 15, n. 4, p. 599-631.

Ore Deposits

Beane, R. E., 1974: Biotite stability in the porphyry copper environment. Econ. Geol., v. 69, p. 241-256.

Arizona, New Mexico, Utah, and British Columbia, Canada

Cheney, E. S., and Jensen, M. L., 1963: Bearing of stable isotopic geology on the origin of the Gas Hills, Wyoming, uranium district. Econ. Geol., v. 61, n. 1, p. 44-71.

Chrismas, L., Baadsgaard, H., Folinsbee, R. E., Fritz, P., Krouse, H. R., and Sasaki, A., 1968: Rb/Sr, S and O isotopic analyses indicating source and date of the pyrometasomatic copper deposits of Craigmont, British Columbia, abs. Econ. Geol., v. 63, n. 6, p. 702-703.

Chrismas, L., Baadsgaard, H., Folinsbee, R. E., Fritz, P., Krouse, H. R., and Sasaki, A., 1968: Rb/Sr, S and O isotopic analyses indicating source and date of contact metasomatic copper deposits, Craigmont, British Columbia. Econ. Geol., v. 64, n. 5, p. 479-488.

Dymond, J., Heath, G. R., and Corliss, J. B., et al., 1973: Elemental and isotopic geochemistry of metalliferous sediments on the east Pacific rise, abs. (in Cordilleran Section, 69th Meeting). Geol. Soc. Am., Abstr., v. 5, n. 1, p. 36.

Dymond, J., Corliss, J. B., Heath, G. R., et al., 1973: Origin of metalliferrous sediments from the Pacific Ocean. Geol. Soc. Am., Bull., v. 84, n. 10, p. 3355-3373.

Engel, A. E., Clayton, R. N., and Epstein, S., 1958: Variations in the isotopic composition of oxygen (and carbon) in the Leadville limestone (Mississippian) of Colorado as a guide to the location and origin of its mineral deposits. Internat. Geol. Congr., 20th, Mexico 1956, Symposium de Exploración, Geoquimica, v. 1, p. 3-20.

Engel, A. E., 1959: Review and evaluation of studies of the O^{18}/O^{16} ratio in mineral deposits, abs. Econ. Geol., v. 54, n. 7, p. 1343.

Folinsbee, R. E., 1970: Rb/Sr, S and O isotopic analyses indicating source and date of contact metasomatic copper deposits, Craigmont, British Columbia, Canada, A reply to 1970 discussion by L. Chrismas and others of 1969 paper. Econ. Geol., v. 65, n. 1, p. 63-64.

Hall, W. E., and Friedman, I., 1969: Oxygen and carbon isotopic composition of ore and host rock of selected Mississippi Valley deposits (in Geological Survey Research, 1969). U.S. Geol. Survey, Prof. Paper 650-C, p. C140-148.

Harzer, D., and Pilot, J., 1969: Isotopengeochemische untersuchungen an ganglagerstatten des Harzes (Isotope geochemical studies of ore veins in the Harz Mountains). Deut. Ges. Geol. Wiss., Ber., Reihe B, Mineral. Lagerstattenforsch, v. 14, n. 2, p. 129-138.

Herrick, D. C., Rose, A. W., and Deines, P., 1972: Mineralological and isotopic studies of the Cornwall iron deposit, Pennsylvania, abs. Geol. Soc. Am., Abstr., v. 4, n. 7, p. 534.

Herrick, D. C., 1974: An isotopic study of the magnetite-chalcopyrite deposit at Cornwall, Pennsylvania, abs. Dissert. Abstr. Internat., v. 34, n. 8, p. 3841B.

Heyl, A. V., Landis, G. P., and Zartman, R. E., 1974: Isotopic evidence for the origin of Mississippi Valley-type mineral deposits, a review (in Stable Isotopes as Applied to Problems of Ore Deposits). Econ. Geol., v. 69, n. 6, p. 992-1006.

Hoekstra, H. R., and Katz, J. J., 1956: Isotope geology of some uranium minerals, proceedings of the international conference on the peaceful uses of atomic energy, August 1955, Geneva. U.S. Geol. Survey, Prof. Paper 300, p. 543-547.

Hoekstra, H. R., 1956: Oxygen isotope variations in some uranium minerals (in Nuclear Processes in Geologic Settings). Natl. Acad. Sci., Natl. Res. Coun. Pub. 400, p. 160-165.

Kiss, J., and Cornicles, I., 1972: Origin of the ore veins in Matra Mountain after isotope investigations (in International Symposium on the Mineral Deposits of the Alps, 2nd Proceedings). Geol. Razprave Porocila, v. 15, p. 383-390.

Kolodny, Y.,1970: Studies in geochemistry of uranium and phosphorites, abs. Dissert. Abstr. Internat., v. 30, n. 11, p. 5103B.

Linares, E., 1966: Geologia isotopica de yacimiento Huemul provincia de Mendoza (Isotopic geology of the Huemel mine, Mendoza Province), English abstract. Assoc. Geol. Argentina Rev., v. 21, n. 3, p. 181-189.

Lovering, T. S., et al., 1963: Significance of O^{18}/O^{16} and C^{13}/C^{12} ratios in hydrothermally dolomitized limestones and manganese carbonate replacement ores of the Drum Mountains, Juab County, Utah. U.S. Geol. Survey, Prof. Paper 475-B, p. 1-9.

Lovering, T. S., McCarthy, J. H., and Friedman, I., 1964: Significance of O^{18}/O^{16} and C^{13}/C^{12} ratios in hydrothermally dolomitized limestones and manganese carbonates replacement ore of the Drum Mountains, Juab County (in Chemistry of the Earth's Crust, v. 2). Jerusalem, Israel Program for Scientific Translations, p. 658-671.

Marowsky, G., 1969: Schwefel-, kohlenstoff- und sauerstoff-isotopenuntersuchungen am kupferschiefer als beitrag zur genetischen deutung (Sulfur, carbon and oxygen isotope studies on the Permian "Kupferschiefer"), English summary. Contrib. Mineral. Petrol., Beitr. Mineral. Petrol., v. 22, n. 4, p. 290-334.

Mel'nik, Yu. P., and Lugovaya, I. P., 1972: O proiskhozhdenii rudnykh mineralov dokembriyskikh zhelezistykh kvartsitov po dannym izucheniya izotopnogo sostava kisloroda (Genesis of ore mineral in Precambrian ferruginous quartzites, based on the isotope composition of oxygen data), English summary. Geokhim. Akad. Nauk SSSR, n. 10, p. 1215-1226.

Ukraine, USSR

Ohmoto, H., and Rye, R. O., 1970: The Bluebell mine of British Columbia, pt. 1 Mineralogy, paragenesis, fluid inclusions, and the isotopes of hydrogen, oxygen and carbon. Econ. Geol., v. 65, n. 4, p. 417-437.

O'Neil, J. R., and Silberman, M. L., 1973: Stable isotopes relations in epithermal Au-Ag deposits, western USA, abs. Geol. Soc. Am., Abstr., v. 5, n. 7, p. 758.

Pease, M. H., Jr., 1962: Some characteristics of copper mineralization in Puerto Rico (in 3rd Caribbean Geol. Conf., 1962, Kingston, Jamaica, trans.). Jamaica Geol. Surv., pub. 95, p. 107-112.

Perry, E. C., Jr., Tan, F. C., and Morey, G. B., 1972: Stable isotope geochemistry of the Biwabik iron formation (in Precambrian Iron Formations of the World). Econ. Geol., v. 68, n. 7, p. 1110-1125.

Robinson, B. W., 1974: The origin of mineralization at the Tui mine, Te Aroha, New Zealand, in the light of stable isotope studies (in Stable Isotopes as Applied to Problems of Ore Deposits). Econ. Geol., v. 69, n. 6, p. 910-925.

Robinson, B. W., and Badham, J. P. N., 1974: Stable isotope geochemistry and the origin of the Great Bear Lake silver deposits, Northwest Territories, Canada. Can. Jour. Earth Sci., v. 11, n. 5, p. 698-711.

Rye, D., and Rye, R. O., 1972: The origin of the Homestake gold deposit, South Dakota, in the light of stable isotope studies. Geol. Soc. Am., Abstr., v. 4, n. 7.

Rye, R. O., and Hall, W. E., 1971: Carbon, hydrogen, oxygen and sulfur isotope study of the Darwin silver-lead-zinc deposits, southern California. Econ. Geol., v. 69, p. 468-481.

Sawkins, F. J., and Rye, D. M., 1974: Relationship of Homestake-type gold deposits to iron-rich Precambrian sedimentary rocks. Inst. Min. Metall., Trans., Sect. B, v. 83, n. 810, p. B56-B59.

Schau, M. P., 1970: Rb/Sr, S, O isotopic analyses indicating source and date of contact metasomatic copper deposits, Craigmont, British Columbia, Canada; discussion of paper by L. Chrismas and others, 1969. Econ. Geol., v. 65, n. 1, p. 62-63.

Scherp. A., and Stadler, G., 1973: Aspekte der erzbildung im Siegerland (Aspects of ore genesis in Siegerland, Germany), English summary. Dtsch. Geol. Ges., Z., v. 124, pt. 1, p. 51-59.

Sheppard, S. M. F., Nielson, R. L., and Taylor, H. P., Jr., 1967: Hydrogen and oxygen isotope variations in minerals from porphyry copper deposits. Econ. Geol., v. 62, n. 6, p. 875; and Geol. Soc. Am. Spec. Paper 115, p. 203.

New Mexico, Nevada, Montana, Chili, El Salvador, Peru and Puerto Rico

Sheppard, S. M. F., Nielsen, R. L., and Taylor, H. P., Jr., 1969: Oxygen and hydrogen isotope ratios of clay minerals from porphyry copper deposits. Econ. Geol., v. 64, n. 7, p. 755-777.

New Mexico, Nevada, Montana, Chile, Peru and Puerto Rico

Sheppard, S. M. F., Nielsen, R. L., and Taylor, H. P., Jr., 1971: Hydrogen and oxygen isotope ratios in minerals from porphyry copper deposits. Econ. Geol., v. 66, n. 4, p. 515-542.

Solomon, M., Rafter, T. A., and Jensen, M. L., 1969: Isotope studies on the Rosebery, Mount Farrell and Mount Lyell ores, Tasmania. Miner. Deposita, v. 4, n. 2, p. 172-179.

Solomon, M., Rafter, T. A., and Dunham, K. C., 1971: Sulfur and oxygen isotope studies in the northern Pennines in relation to ore genesis. Inst. Mining Met., Trans., v. 80, Sect. B, n. 777, p. B259-B275.

Solomon, M., Rafter, T. A., and Dunham, K., 1972: Sulfur and oxygen isotope studies in the northern Pennines in relation to ore genesis; discussion. Inst. Mining Met., Trans., Sect. B, v. 81, n. 789, p. B172-B177.

Solomon, M., Rafter, T. A., and Dunham, K., 1973: Sulfur and oxygen isotope studies in the northern Pennines in relation to ore genesis. Inst. Mining Met., Trans., Sect. B, v. 82, n. 795, p. B46.

Taylor, H. P., Jr., 1974: The application of oxygen and hydrogen isotope studies to problems of hydrothermal alteration and ore deposition (in Stable Isotopes as Applied to Problems of Ore Deposits). Econ. Geol., v. 69, n. 6, p. 843-883.

Wedepohl, K. H., 1971: "Kupferschiefer" as a prototype of syngenetic sedimentary ore deposits (in International Association of the Genesis of Ore Deposits, Tokyo-Kyoto, Meetings, Papers and Proceedings). Soc. Min. Geol. Japan, Spec. Issue, n. 3, p. 268-273.

Wolf, D., and Espovo, E., 1972: Zur geochemie bolivianischer kassiterite (The geochemistry of Bolivian cassiterites), Russian, English and Spanish summaries. Z. Angew Geol., v. 18, n. 10, p. 459-468.

G. ROCKS AND OCEAN WATER

Garlick, G. D., and Dymond, J. R., 1970: Oxygen isotope exchange between volcanic materials and ocean water. Geol. Soc. Am., Bull., v. 81, n. 7, p. 2137-2141.

Moore, J. G., 1970: Water content of basalt erupted on the ocean floor. Contrib. Mineral. Petrol., Beitr. Mineral. Petrol., v. 28, n. 4, p. 272-279.

Muehlenbachs, K., and Clayton, R. N., 1970: O^{18}/O^{16} studies of fresh and weathered submarine basalts. EOS, Am. Geophys. Union, Trans., v. 51, n. 4, p. 444.

Muehlenbachs, K., and Clayton, R. N., 1971: The oxygen isotope geochemistry of submarine greenstone, abs. EOS, Am. Geophys. Union, Trans., v. 52, n. 4, p. 365.

Muehlenbachs, K., and Clayton, R. N., 1971: Oxygen isotope ratios of submarine diorites and their constituent minerals. Can. Jour. Earth Sci., v. 8, n. 12, p. 1591-1595.

Muehlenbachs, K., and Clayton, R. N., 1972: Oxygen isotope studies of fresh and weathered submarine basalts. Can. Jour. Earth Sci., v. 9, n. 2, p. 172-184.

Muehlenbachs, K., and Clayton, R. N., 1972: Oxygen isotope geochemistry of submarine greenstones. Can. Jour. Earth Sci., v. 9, n. 5, p. 471-478.

Yeh, H.-W., and Savin, S. M., 1972: Rate of oxygen isotope exchange between clay minerals and sea water, abs. Clay Mineral Conf., Program Abstr., n. 21, p. 26.

12. HYDROLOGIC CYCLE

A. GENERAL

Arnason, B., and Sigurgeirsson, Th., 1967: Hydrogen isotopes in hydrological studies in Iceland (in Isotopes in Hydrology, Proceedings, Symposium of International Atomic Energy Agency and International Union Geodesy and Geophysics, November 14-18, 1966, Vienna). Vienna, IAEA, STI-PUB/141, paper n. SM 83/3, p. 35-47.

Bowen, R., 1969: Progress in isotopic hydrology; the combined environmental isotope approach. Sci. Progr., v. 57, n. 228, p. 559-575.

Brown, R. M., 1970: Distribution of hydrogen isotopes in Canadian waters (in Isotope Hydrology, Proceedings, Symposium of International Atomic Energy Agency and UNESCO, March 9-13, 1970, Vienna). Vienna, IAEA, STI/PUB/255, paper n. SM-120/1, p. 3-21.

Brown, R. M., Robertson, E., and Thurston, W. B., 1971: Deuterium content of Canadian waters; II. Chalk River, Ontario. Atomic Energy Canada Ltd. (NTIS), 60 p.

Conrad, G., and Fontes, J.-C., 1970: Isotopic hydrological studies of the northwest Sahara (in Isotope Hydrology Proceedings, Symposium of International Atomic Energy Agency and UNESCO, March 9-13, 1970, Vienna). Vienna, IAEA, STI/PUB/255, Paper n. SM-129/24, p. 405-419.

Craig, H., 1961: Isotopic variations in meteoric waters. Science, v. 133, n. 3465, p. 1702-1703.

Cziki, K., and Fodor, J., 1956: Study of the deuterium content of natural inland waters and vegetable saps. Acta Geol. Acad. Sci. Hungaricae, tomus 4, fasc. 2, p. 131-142.

Dincer, T., Martinec, J., Payne, B. R., and Yen, C. K., 1970: Variations of the tritium and oxygen-18 content in precipitation and snowpack in a representative basin in Czechoslovakia (in Isotope Hydrology, Proceedings, International Atomic Energy Agency and UNESCO, March 9-13, 1970, Vienna). Vienna, IAEA, STI/PUB-255, Paper n. SM-129/3, p. 23-42.

Dincer, T., and Payne, B. R., 1971: An environmental isotope study of the southwestern karst region of Turkey. J. Hydrol., v. 14, n. 3/4, p. 233-258.

Eichler, R., 1966: Deuterium-isotopengeochemie des grund- und oberflächenwassers (Deuterium isotope geochemistry of ground and surface water), English, French and Russian summaries. Geol. Rundsch., v. 55, n. 1, p. 144-159.

Epstein, S., and Mayeda, T., 1953: Variations of O^{18} content of waters from natural sources. Geochim. Cosmochim. Acta, v. 4, n. 5, p. 213-224.

Greenland, Great Lakes, Alaska, California currents

Epstein, S., 1956: Variation of the O^{18}/O^{16} ratios of fresh waters and ice (in Nuclear Processes in Geologic Settings). Natl. Acad. Sci., Natl. Res. Coun. Pub. 400, p. 20-28.

Oregon, California, Alaska, Alberta, Greenland

Eriksson, E., and Bolin, B., 1965: Oxygen-18, deuterium and tritium in natural waters and their relations to the global circulation of water (in Radioactive Fallout from Nuclear Weapons Tests). USAEC Proceedings, 2nd Conference, November 3-6, 1964, Germantown, Maryland.

Eriksson, E., 1965: Deuterium and oxygen-18 in precipitation and other natural waters--some theoretical considerations. Tellus, v. 17, n. 4, p. 498-512.

Faure, H., Fontes, J. Ch., Gischler, C. E., et al., 1970: Un example d'étude d'hydrogéologie isotopique en pays sémi-aride, le bassin du Lac Tchad (An example of an isotopic hydrogeologic study in semi-arid lands, the Lake Chad basin), English summary. J. Hydrol., v. 10, n. 2, p. 141-150.

Friedman, I., 1953: Deuterium content of natural waters and other substances. Geochim. Cosmochim. Acta, v. 4, n. 1, p. 89-103.

Water from United States; fumarolic gases from Yellowstone National Park

Friedman, I., 1957: Determination of deuterium-hydrogen ratios in Hawaiian waters. Tellus, v. 9, n. 4, p. 553-556.

Friedman, I., and Redfield, A. C., 1961: Applications to hydrology of deuterium variation in natural waters, abs. J. Geophys. Res., v. 66, n. 8, p. 2530.

Friedman, I., Redfield, A. C., Schoen, B., and Harris, J., 1963: The variation of the deuterium content of natural waters in the hydrologic cycle. J. Geophys. Res., v. 68, n. 13, p. 3781.

Friedman, I., Sigurgeirsson, T., and Gardarsson, O., 1963: Deuterium in Iceland waters. Geochim. Cosmochim. Acta, v. 27, n. 6, p. 553-561.

Friedman, I., Redfield, A. C., Schoen, B., and Harris, J., 1964: The variation of the deuterium content of natural waters in the hydrologic cycle. Rev. Geophys., v. 2, n. 1, p. 177-224.

Gat, J. R., and Dansgaard, W., 1972: Stable isotope survey of the fresh water occurrences in Israel and the northern Jordan rift valley. J. Hydrol., v. 16, n. 3, p. 177-211.

Gorbushina, L. V., Vetshteyn, G. A., Malyuk, Ye., Iosifova, V., and Arutyunyants, R. R., 1972: Isotopic composition of oxygen and hydrogen in sulfide waters of the Sochi-Adler artesian basin (Izotopnyy sostav kisloroda i vodorada sul'fidnykh vod sochi-Alderskogo artezianskogo basseyna). Geokhimiya, v. 9, n. 5, p. 1102-1106.

USSR

Hacker, P., 1973: Ergebnisse hydrologischer untersuchungen und messungen der umweltisotope im einzugsgebiet des Passaiter beckens, Mittelsteiermark (Hydrologic studies and measurements of the natural isotopes in the Passail basin region, central Styria, Austria), French summary. Steirische Beitr. Hydrogeol., v. 25, p. 139-182.

Hagemann, R., Nief, G., and Roth, E., 1970: Absolute isotopic scale for deuterium analysis of natural waters, absolute D/H ratio for SMOW. Tellus, v. 22, n. 6, p. 712-715.

Harzer, D., Pilot, J., and Rösler, H. J., 1968: Sauerstoff-isotopenuntersuchungen am oberflächen- und niederschlags-wassern des Harzes (Oxygen isotopic investigations of surface and meteoric waters of the Harz region), English and Russian summaries. Bergakademie, v. 20, n. 6, p. 329-332.

Hitchon, B., and Friedman, I., 1969: Geochemistry and origin of formation waters in the western Canada sedimentary basin; pt. 1, stable isotopes of hydrogen and oxygen. Geochim. Cosmochim. Acta, v. 33, n. 11, p. 1321-1349.

Hitchon, B., and Krouse, H. R., 1972: Hydrogeochemistry of surface waters of the MacKenzie River drainage basin, Canada; III., stable isotopes of oxygen, carbon and sulfur. Geochim. Cosmochim. Acta, v. 36, n. 12, p. 1337-1357.

Horibe, Y., and Kobayakawa, M., 1960: Deuterium abundance of natural waters. Geochim. Cosmochim. Acta, v. 20, n. 3/4, p. 273-283.

Japan, Pacific and Antarctic Oceans

Issar, A., Rosenthal, E., Eckstein, Y., and Bogoch, R., 1971: Formation waters, hot springs and mineralization phenomena along the eastern shore of the Gulf of Suez. Bull. Internat. Assoc. Sci. Hydrol., v. 16, n. 3, p. 25-44.

Jacobshagen, V., 1961: Die isotopenzusammen setzung natürlicher wässer und ihre änderungen im wasserkreislauf (The isotopic composition of natural water and its changes during the water cycle). Geol. Rundschau, v. 51, n. 1, p. 281-290.

Judy, C. H., 1972: Heavy water--a natural tracer. Indiana Acad. Sci., Proc., v. 81, p. 242-245.

Kraynova, L. P., 1964: Kizucheniyu izotopnogo sostava vod vysokogornykh istachnikov (On the study of the isotopic composition of waters from high mountain sources). Akad. Nauk Uzbek, SSR Uzbek. Geol. Zhur., n. 6, p. 83-85.

Krouse, H. R., 1967: General comment on microbiological activity and its effects on isotopic studies of water (in Isotope Techniques in the Hydrologic Cycle, Univ. Illinois 1965 Symposium). Am. Geophys. Union, Geophys. Mon. Ser., n. 11, p. 195.

Krouse, H. R., and Sasaki, A., 1969: Isotopic studies of springs in western north Africa, abs. Geol. Soc. Am. Spec. Paper 121, p. 167.

Krouse, H. R., and Mackay, J. R., 1971: Application of H_2O^{18}/H_2O^{16} abundances to the problem of lateral mixing in the Liard-Mackenzie river system. Can. Jour. Earth Sci., v. 8, n. 9, p. 1107-1109.

Leal, J. M., Salat, F., Campos, M. M., et al., 1973: Caracterizacao de aguas do nordeste Brasileiro com isotopos ambientais (The characterization of the waters of northeastern Brazil with natural isotopes), abs. Resumo das communicacoes simposios e conferencias, Simposio de Hidrogeologia do Nordeste, Congr. Bras. Geol., n. 27, Bol. 2, p. 80-83.

Mangano, F., Marce, A., Meybeck, M., et al., 1970: Idologia isotopica metodologia e prime applicazioni alle sorgenti Fiume e Bella, Madonie orientali (Isotope hydrology, methodology and primary application to the sources at Fiume and Bella, eastern Madonie Mountains), French summary. Riv. Min. Sicil., v. 21, n. 124-126, p. 223-231.

Sicily

Millson, M. F., 1971: Hydrogen isotope ratios in a recycling system. J. Water Pollution Control Fed., v. 43, n. 11, p. 2287-2290.

Miyake, Y., and Matsuo, S., 1962: A note on the deuterium content in the atmosphere and the hydrosphere. Pap. Meteor. Geophys., v. 13, n. 3-4, p. 245-259.

Molochnova, V. A., Sokolov, M. M., and Gorev, A. V., 1967: Deuterium content in natural waters. Geochem. Internat., v. 4, n. 3, p. 484-489.

Mook, W. G., 1970: Stable carbon and oxygen isotopes of natural waters in the Netherlands (in Isotope Hydrology). International Atomic Energy Agency, Vienna, IAEA, p. 163-190.

Morgante, S., Mosetti, F., and Tongiori, E., 1966: Moderne indagini idrologiche nella zona di Gorizia (Recent hydrological investigations in the Gorizia zone), English summary. Boll. Geofisica Teor. ed Appl., v. 8, n. 30, p. 114-137.

Moser, H., and Stichler, W., 1971: Die verwendung des deuterium- und sauerstoff-18-gehalts bei hydrologischen untersuchungen (The utilization of deuterium and oxygen-18 concentrations in hydrological investigations). Geol. Bavarica, n. 64, p. 7-35.

Moser, H., Stichler, W., and Trimborn, P., 1971: Deuterium-und sauerstoff-isotopengehalt von oberflächenwässern des bayerischen alpervorlandes. Besondere Mitt. z. Deutsch. Gewässerkundlichen Jb. 35, p. 445-456a, GSF-R 55.

Moser, H., 1972: Verwendung des deuterium-und sauerstoff-18-gehalts bei hydrologischen untersuchungen. Gas, Wasser, Abwasser 52, p. 329-334, GSF-R 65.

Moser, H., Stichler, W., and Trimborn, P., 1973: Deuterium and oxygen-18 measurements on surface waters of the Bavarian prealps (in Symposium on Hydrometry, Koblenz, West Germany, September 1970). Internat. Assoc. Hydrol. Sci. Pub. n. 99, v. II, p. 615-626.

Pilot, J., and Vogel, J., 1972: Isotopengeochemische untersuchungen von waessern (Isotope geochemical analyses of waters). Dtsch. Ges. Geol. Wiss. Ber., Reihe A, Geol. Palaontol., v. 17, n. 2, p. 233-247.

Polyakov, Yu. A., 1964: Use of deuterium to study the movement of surface and underground water. Pochvovedeniye, Soviet Soil Sci., Trans. n. 11, p. 1090-1095.

Porras, M. J., 1973: Estudio hidrogeologico de la cuenca del Duero; situacion de los conocimientos hidrogeologicas basicos (A hydrogeologic study of the Duero basin, Spain, basic data. Spain Inst. Geol. Min. Bol. Geol. Min., v. 84, n. 5, p. 336-346.

Redfield, A. C., and Friedman, I., 1969: The effect of meteoric water, melt water and brine on the composition of polar sea water and of the deep waters of the ocean. Pergamon Press, Deep Sea Res., Supplement v. 16, p. 197-214.

Baffin Bay, Arctic Ocean

Schwarcz, H. P., and Cortecci, G., 1974: Isotopic analyses of spring and stream water sulfate from the Italian Alps and Apennines. Chem. Geol., v. 13, n. 4, p. 285-294.

Siegenthaler, U., Oeschger, H., and Tongiorgi, E., 1970: Tritium and oxygen-18 in natural water samples from Switzerland (in Isotope Hydrology, Proceedings Symposium of International Atomic Energy Agency and UNESCO, March 9-13, 1970). Vienna, IAEA, STI/PUB/255, Paper n. SM-129/22, p. 373-385.

Soyfer, V. N., Brezgunov, V. S., and Vlasova, L. S., 1967: Role of stable hydrogen isotopes in the study of geological processes. Geochem. Internat., v. 4, n. 3, p. 490-497.

USSR

Tan, F. C., Pearson, G. J., and Walker, R. W., 1974: O^{18}/O^{16} ratios of water masses in the Gulf of St. Lawrence, Canada. EOS, Am. Geophys. Union, Trans., v. 55, n. 4, p. 315.

Thatcher, L. L., 1967: Water tracing in the hydrologic cycle (in Isotope Techniques in the Hydrologic Cycle, Univ. Illinois, Symposium, 1965). Am. Geophys. Union, Geophys. Mon. Ser., n. 11, p. 97-108.

Tyminskiy, V. G., Sultankhodzhayev, A. N., Taneyev, R. N., et al., 1971: Kompleksnoye ispol'zovaniye izotopn kh pokazatelev pri reshenii zadach paleogidrogeologii (Multiple utilization of isotopic indicators for solving paleohydrogeologic problems). Mosk. Ova. Ispyt. Prir. Byull. Otd. Geol., v. 46, n. 4, p. 130-138.

Uklonskiy, A. S., 1953: Predvaritel'nyye issledovaniya izotopnogo slstava poverkhnostnykh i podzemnykh vod Uzbekistana (Preliminary investigations of the isotopic composition of surface and ground water of Uzbekistan). Akad. Nauk Uzbek. SSR, Zapiski Uzbekistan. Otdel. Ves. Mineral. Obshch., n. 6, p. 47-62.

USSR, Arctic and Antarctic Oceans

Uklonskiy, A. S., Bugayev, V. A., and Glushchenko, V. M., 1966: K voprosu ob izuchenii summarnogo izotopnogo sostava vod Arktiki (On the problem of studies of the total isotopic content of Arctic waters). Akad. Nauk Uzbek. SSR Doklady, n. 6, p. 39-40.

Uklonskiy, A. S., Popov, V. I., Badalov, S. T., et al., 1973: Summarnyy izotopnyy sostav i gidrokhimicheskiye osobennosti prirodnykh vod severo-zapadnoy i severnoy-Fergany (Total isotope composition and hydrochemical characteristics of natural waters in northeastern and northern Fergana), Uzbek. summary. Uzbek. Geol. Zh., n. 1, p. 60-63.

Vetshteyn, V. E., Gulyanitskaya, T. P., Soyner, V. N., et al., 1967: Izotopnyy sostav kisloroda i vodoroda vod otkrytykh vodoyemov i vulkanogennykh istochnikov (Isotopic composition of oxygen and hydrogen in the waters of surface reservoirs and volcanic springs). Geokhim., n. 6, p. 737-739.

Vetshteyn, V. E., Gulyanitskaya, T. P., Soyfer, V. N., et al., 1968: Isotopic composition of oxygen and hydrogen in waters of open basins and volcanic springs. Geochim. Int., v. 4, n. 3, p. 601-604.

Vogel, J. C., Ehhalt, D., and Roether, W., 1963: A survey of the natural isotopes of water in South Africa (<u>in</u> Radioisotopes in Hydrology), French, Russian and Spanish summaries. Vienna, IAEA, p. 407-415.

White, D. F., Barnes, I., and O'Neil, J. R., 1971: Spring water of nonmeteoric origin, California Coast Ranges, abs. Geol. Soc. Am., Abstr., v. 3, n. 7, p. 750.

Zimmerman, U., and Zoetl, J., 1971: Deuterium-und sauerstoff-18-gehalt von gasteiner thermal-und kaltwaessern (Deuterium and oxygen-18 content of thermal and fresh waters of the Bod Gastein area), English summary. Steirische Beitr. Hydrogeol., v. 23, p. 127-132.

B. METEORIC WATER-ROCK REACTIONS

Addy, S. K., and Ypma, P. J. M., 1973: Metamorphism of sulfide deposits and the metamorphogenic circulation of pore water at Ducktown, Tennessee, abs. Geol. Soc. Am., Abstr., v. 5, n. 7, p. 529-530.

Giletti, B. J., 1973: Constraints on the amount of meteoric water which passed through an epizonal intrusive and lowered its O^{18} content, abs. (in Fall Annual Meeting, San Francisco, 1973, Section of Volcanology, Geochemistry, and Petrology; Isotope and Low-Temperature Geochemistry). EOS, Am. Geophys. Union, Trans., v. 54, n. 11, p. 1227-1228.

Sheppard, S. M. F., and Taylor, H. P., Jr., 1974: Hydrogen and oxygen isotope evidence for the origin of water in the boulder batholith and the Butte ore deposits, Montana (in Stable Isotopes as Applied to Problems of Ore Deposits). Econ. Geol., v. 69, n. 6, p. 926-946.

Taylor, B. E., 1974: Communication between magmatic and meteoric fluids during formation of Fe-rich skarns in north central Nevada, abs. EOS, Am. Geophys. Union, Trans., v. 55, n. 4, p. 478.

Taylor, H. P., Jr., and Epstein, S., 1969: Hydrogen isotope evidence for influx of meteoric ground water into shallow igneous intrusions, abs. Geol. Soc. Am., Spec. Paper 121, p. 294.

Taylor, H. P., Jr., 1970: Oxygen isotope evidence for large-scale interaction between meteoric ground waters and tertiary diorite intrusions, western Cascade range, Oregon, abs. EOS, Am. Geophys. Res., v. 76, n. 32, p. 7855-7874.

Taylor, H. P., Jr., and Forester, R. W., 1971: Low O^{18} igneous rocks from the intrusive complexes of Skye, Mull and Ardnamurchan, western Scotland. J. Petrol., v. 12, n. 3, p. 465-497.

C. ATMOSPHERE

Begemann, F., and Friedman, I., 1959: Tritium and deuterium content of atmospheric hydrogen. Zeitschr. Naturforschung., v. 14a, n. 12, p. 1024-1031.

Buffalo, New York, and Hamburg, Germany

Begemann, F., and Friedman, I., 1968: Tritium and deuterium in atmospheric methane. J. Geophys. Res., v. 73, n. 4, p. 1149-1153.

Gary, Indiana, and Oberhauser, Germany

Begemann, F., and Friedman, I., 1968: Isotopic composition of atmospheric hydrogen. J. Geophys. Res., v. 73, n. 4, p. 1139-1147.

Munich, Germany

Chevallier, H., Chretien, R., and Letolle, R., 1971: Les corbicules du gisement pleistocene de Cergy (Corbicula from the Pleistocene of Cergy) (in Etudes sur la Quaternaire dans le Monde, v. 1). Assoc. Fr. Etude Quat., Bull., Suppl. n. 4, p. 425-431.

Craig, H., and Gordon, L. I., 1965: Isotopic oceanography-deuterium and oxygen-18 variations in the ocean and the marine atmosphere (in Symposium on Marine Geochemistry, 1964). Rhode Island Univ., Narragansett Marine Lab., Occasional Pub. 3-1965, p. 277-374.

Craig, H., and Gordon, L. I., 1967: Deuterium and oxygen-18 variations in the ocean and the marine atmosphere. Consiglio Nazionale Delle Richerche Laboratorio di Geologia Nucleare Pisa (NTIS), p. 1-122.

Craig, H., and Begemann, F., 1970: Oxygen-18 variations in atmospheric carbon dioxide--The HASL data. J. Geophys. Res., v. 75, n. 9, p. 1723-1726.

Dole, M., Rudd, D. P., Lane, G. A., Zaukelies, D. A., and Brown, J. B., 1954: The height and geographic distribution of the oxygen isotopes. Northwestern Univ., Dept. of Chemistry, Contract AF(122)-157 and AF(604)-587, Final report, 108 p.

U.S., South America, Central America, Finland, Holland, Japan, Australia and Antarctica

Dole, M., Lane, G. A., Rudd, D. P., and Zaukelies, D. A., 1954: Isotopic composition of atmospheric oxygen and nitrogen. Geochim. Cosmochim. Acta, v. 6, n. 2/3, p. 65-78.

Pacific Ocean

Dole, M., 1956: The oxygen isotope cycle in nature (<u>in</u> Nuclear Processes in Geologic Settings). Natl. Acad. Sci., Natl. Res. Coun., Pub. 400, p. 13-19.

Donohue, T. M., 1969: Deuterium in the upper atmosphere of Venus and Earth. J. Geophys. Res., v. 74, n. 5, p. 1128-1137.

Edwards, G., 1955: Isotopic composition of meteoric hydrogen. Nature, v. 176, n. 4472, p. 109-111.

Ehhalt, D., Israel, G., Roether, W., et al., 1963: Tritium and deuterium content of atmospheric hydrogen. J. Geophys. Res., v. 68, n. 13, p. 3747-3751.

Ehhalt, D., 1966: Tritium and deuterium in atmospheric hydrogen. Tellus, v. 18, n. 2/3, p. 249-255.

Friedman, I., 1974: Isotope composition of atmospheric hydrogen, 1967-1969. J. Geophys. Res., v. 79, n. 6, p. 785-788.

Colorado

Gat, J. R., and Carmi, I., 1970: Evolution of the isotopic composition of atmospheric waters in the Mediterranean Sea area. J. Geophys. Res., v. 75, n. 15, p. 3039-3048.

Gonsior, B., and Friedman, I., 1962: Tritium and deuterium im atmosphärischen wasserstoff (Tritium and deuterium in atmospheric hydrogen), English abstract. Zeitschr. Naturforschung., v. 17a, n. 12, p. 1088-1091.

Hamburg, Germany

Gonsior, B., Friedman, I., and Ehhalt, D., 1963: Measurements of tritium and deuterium concentration in atmospheric hydrogen. J. Geophys. Res., v. 68, n. 13, p. 3753-3756.

Gonsior, B., Friedman, I., and Lindenmayr, G., 1965: New tritium and deuterium measurements in atmospheric hydrogen. Tellus, v. 18, n. 2/3, p. 256-261.

Horibe, Y., Shigehara, K., and Takakuwa, Y., 1973: Isotope separation factor of carbon dioxide-water system and isotopic composition of atmospheric oxygen. J. Geophys. Res., v. 78, n. 15, p. 2625-2629.

Isono, K., Kombayasi, M., and Takahashi, T., 1966: Physical study of solid precipitation from convective clouds over the sea; pt. 1, deuterium content of snow crystals with respect to crystal shapes and their relation to origins of the water vapor of snowfall. Meteor. Soc. Japan J., Ser. 2, v. 44, n. 3, p. 178-184.

Japan and Japan Sea

Keeling, C. D., 1958: The concentration and isotopic abundance of atmospheric carbon dioxide in rural areas. Geochim. Cosmochim. Acta, v. 13, n. 4, p. 322-334.

Pacific Coast of North America

Keeling, C. D., 1961: The concentration and isotopic abundance of carbon dioxide in rural and marine air. Geochim. Cosmochim. Acta, v. 24, n. 3/4, p. 277-298.

Eastern Pacific Ocean

Kockarts, G., and Nicolet, M., 1962: Le problème aéronomique de l'hélium et de l'hydrogène neutres (The aeronometric problem of neutral helium and hydrogen), English and Russian abstracts. Annales Géophysiques, v. 18, n. 3, p. 269-290.

Kroopnick, P. M., and Craig, H., 1971: Isotopic composition of molecular oxygen in the atmosphere and the sea, abs. EOS, Am. Geophys. Union, Trans., v. 52, n. 4, p. 255.

Kroopnick, P. M., and Craig, H., 1972: Atmospheric oxygen, isotopic composition and solubility fractionation. Science v. 175, n. 4017, p. 54-55.

Kulp, J. L., Gilett, B. J., and Broecker, W. S., 1956: Radioactive isotopes in the atmosphere. Columbia Univ., Lamont Geological Observatory, Contract AF 19(604)-851, Final Rpt., 41 p.

Layton, B. R., 1969: Tritium in atmospheric hydrogen and atmospheric water. Fayetteville, Arkansas Univ., Thesis, (NTIS), 62 p.

Steven, C. W., Krout, L., Walling, D., Venters, A., Engelkemeir, A., and Ross, L. E., 1972: The isotopic composition of atmospheric carbon monoxide. Earth Planet. Sci. Lett., v. 16, n. 2, p. 147-165.

Stewart, J. B., 1963: Symposium on trace gases and natural and artificial radioactivity in the atmosphere. London, Meteor. Mag., v. 92, n. 1094, p. 279-281.

Teegarden, B. J., 1967: Cosmic ray production of deuterium and tritium in the earth's atmosphere. J. Geophys. Res., v. 72, n. 19, p. 4863-4868.

Vinogradov, A. P., et al. : Isotope fractionation of atmospheric oxygen. Geokhimiya, n. 3, p. 241-253.

D. OCEANS

General

Babinets, A. Ye., Vetshteyn, V. Ye., and Davidyuk, L. A., 1967: Deyaki rezhul'taty vyvchennya vmistu izhotopu O^{18} u vodakh tropichnoyi Atlantyky (Some results of study of the O^{18} content in waters of the tropical Atlantic), English and Russian summaries. Akad. Ukrayin. RSR Dopovidi, Ser. B, n. 3, p. 230-232.

Chappell, J., 1974: Relationships between sea levels, O^{18} variations and orbital perturbations during the past 250,000 years. Nature, v. 252, n. 5480, p. 199-202.

Chase, C. G., and Perry, E. C., Jr., 1972: The oceans, growth and oxygen isotope evolution. Science, v. 177, n. 4053, p. 992-994.

Chase, C. G., and Perry, E. C., Jr., 1972: Stable isotope tracers and a shrinking ocean, abs. Geol. Soc. Am., Abstr., v. 4, n. 7, p. 469-470.

Craig, H., 1957: Isotopic tracer techniques for measurement of physical processes in the sea and the atmosphere. Washington, D.C., National Res. Coun., Pub. 551, p. 103-120.

Craig, H., and Gordon, L. I., 1966: Isotopic study of the formation of deep water masses, abs. Am. Geophys. Union, Trans., v. 47, n. 1, p. 112.

Craig, H., 1969: Geochemistry and origin of the Red Sea brines (in Hot Brines and Recent Heavy Metal Deposits in the Red Sea). New York, Springer-Verlag.

Craig, H., and Kroopnick, P., 1970: Oxygen-18 variations in dissolved oxygen in the sea, abs. EOS, Am. Geophys. Union, Trans., v. 51, n. 4, p. 325.

Dansgaard, W., 1960: The content of heavy oxygen isotope in the water masses of the Philippine trench. Deep Sea Res., v. 1, n. 4, p. 346-350.

Duursma, E. K., 1972: Geochemical aspects and applications of radionuclides in the sea. Oceanogr. Mar. Biol., Annu. Rev., v. 10, p. 137-223.

Ehhalt, D. H., 1969: On the deuterium salinity relationship in the Baltic Sea. Tellus, v. 21, n. 3, p. 429-435.

Epstein, S., 1962: The oxygen isotope composition of marine waters, abs. J. Geophys. Res., v. 67, n. 9, p. 3555.

Epstein, S., 1967: The stable isotopes. Eng. Sci., v. 31, n. 2, p. 47-51.

Frank, D. J., 1973: Deuterium variations in the Gulf of Mexico and selected organic materials, abs. Dissert. Abstr. Int., v. 33, n. 8, p. 3824B.

Horibe, Y., and Ogura, N., 1968: Deuterium content as a parameter of water mass in the ocean. J. Geophys. Res., v. 73, n. 4, p. 1239-1249.

Horibe, Y., and Shigehara, K., 1973: Oxygen-18 content of dissolved oxygen in the south Pacific, abs. (in Oceanography of the South Pacific). Wellington, New Zealand, Comm., UNESCO, 1972.

Kroopnick, P., 1972: Total CO_2, C^{13}, and dissolved O^{18} at geosecs II in the north Atlantic. Earth Planet. Sci. Lett., v. 16, n. 1, p. 103-110.

Magaritz, M., and Kaufman, A., 1973: Changes in the isotopic composition of east Mediterranean sea water during the Holocene. Nature, v. 244, n. 5408, p. 462-464.

Menache, M., 1966: Variation de la masse volumétrique de l'eau en fontion de sa composition isotopique (Variation of the volumetric mass of water as a function of its isotopic composition). France, Comité Central d'Océanographie et d'Etudes des Côtes, Cahiers Océanographiques, v. 18, n. 6, p. 477-496.

Merlivat, L., and Menache, M., 1973: Etude de profils de deuterium et de tritium en Méditerranée occidentale (Deuterium and tritium profiles in the western Mediterranean Sea), abs. Réun. Annu. Sci. Terre, Programme Résumés, p. 300.

Molcard, R., Cortecci, G., and Noto, P., 1973: Isotopic analysis of the deep staircase structure in the Tyrrhenian Sea. Italy, Saclant ASW Research Centre, La Spezia (NTIS), 13 p.

Nikolayev, S. D., and Popov, S. V., 1973: Primeneniye izotopnogo kislorodnogo metoda k izucheniyu paleogeografii zamknutykh i poluzamknutykh basseynov, tipa Chernago i Kaspiyskogo morey (Applications of the oxygen isotope method to paleogeographic investigations of epicontinental and Mediterranean basins, with the Black and Caspian Seas as examples), abs. Mosk. Ovo. Ispyt. Prir. Byull., Otd. Geol., v. 48, n. 1, p. 158.

Parker, P. L., and Shultz, D., 1970: Variation of O^{18}/O^{16} of dissolved oxygen in marine waters, abs. EOS, Am. Geophys. Union, Trans., v. 51, n. 4, p. 325.

Perry, E. C., Jr., and Chase, C. G., 1972: A model for O^{18} evolution of the ocean, abs. EOS, Am. Geophys. Union, Trans., v. 53, n. 4, p. 555.

Redfield, A. C., and Friedman, I., 1965: Factors affecting the distribution of deuterium in the ocean (in Symposium on Marine Geochemistry, 1964). Rhode Island Univ., Narragansett Marine Lab., Occasional Pub. 3, p. 149-168.

Stahl, W., and Rinow, U., 1973: Sauerstoffisotopenanalysen an mittelmeerwaessern; ein beitrag zur problematik von palaeotemperaturbestimmungen (Oxygen isotope analyses of water samples from the Mediterranean; a contribution on the problems of paleotemperature determination). "Meteor" Forschungsergeb. Reihe C, n. 14, p. 55-59.

Titani, T., Horibe, Y., and Kobayakawa, M., 1961: Deuterium abundance of natural waters in Antarctic, abs. Tokyo, Antarctic Record, n. 11, p. 138-140.

Paleosalinity

Degans, E. T., 1959: Das O^{18}/O^{16} verhältnis im urozean und der geochemische stoffumsatz (The O^{18}/O^{16} ratio in the primordial ocean and the geochemical exchange of matter). Neues. Jahrb. Geologie und Paläontologie Monatsh., n. 4, p. 180-186.

Deuser, W. G., 1972: Late Pleistocene and Holocene history of the Black Sea as indicated by stable isotope studies. J. Geophys. Res., v. 77, n. 6, p. 1071-1077.

Dodd, J. R., and Stanton, R. J., Jr., 1971: Oxygen isotopic determination of paleosalinities within a Pliocene bay, Kettleman Hills, California, abs. Geol. Soc. Am. Abstr., v. 3, n. 4, p. 257-258; and n. 7, p. 574.

Dodd, J. R., and Stanton, R. J., Jr., 1975: Paleosalinites within a Pliocene bay, Kettleman Hills, California; a study of the resolving power of isotopic and faunal techniques. Geol. Soc. Am. Bull., v. 86, p. 51-64.

Duplessy, J. C., and Barbaroux, L., 1973: Répartition des isotopes stables en mil eu éstuarien; application à l'étude isotopique des paléoenvironnements (The distribution of stable isotopes in an estuarine environment; application to paleoenvironments), English summary. Rev. Geogr. Phys. Geol. Dyn., v. 15, n. 5, p. 533-546.

Weber, Dzh. N., 1965: K voprosu ob otnoshenii O-18/O-16 drevnykh okeanov (On the problem of the O-18/O-16 ratios of ancient oceans), English summary. Geokhimiya, n. 6, p. 674-680.

Weber, J. N., 1965: Changes in the oxygen isotopic composition of sea water during the Phanerozoic evolution of the oceans, abs. Geol. Soc. Am. Spec. Paper 82, p. 218-219.

E. INTERSTITIAL WATER

Friedman, I., 1965: Interstitial water from deep sea sediments. J. Geophys. Res., v. 70, n. 16, p. 4066-4067.

Friedman, I., and Hardcastle, K., 1973: Interstitial water studies, leg 15, isotopic composition of water (in Initial Reports of the Deep Sea Drilling Project). Washington, D.C., U.S. Govt. Printing Office, v. 20, p. 901-903.

Friedman, I., and Hardcastle, K., 1974: Deuterium in interstitial waters from Red Sea cores, DSDP leg 23 (in Initial Reports of the Deep Sea Drilling Project). Washington, D.C., U.S. Govt. Printing Office, v. 23, p. 969-970.

Lawrence, J. R., 1973: Interstitial water studies, leg 15 stable oxygen and carbon isotope variations in water, carbonates and silicates from the Venezuela basin, site 149, and the Aues basin, site 148 (in Initial Reports of the Deep Sea Drilling Project). Washington, D.C., U.S. Govt. Printing Office, v. 20, p. 891-899.

Lawrence, J. R., 1973: Depleted O^{18}/O^{16} in DSDP pore waters, abs. Geol. Soc. Am., Abstr., v. 5, n. 7, p. 710.

Lawrence, J. R., 1974: Stable oxygen and carbon isotope variations in the pore waters, carbonates and silicates, site 225 and 228, Red Sea, DSDP leg 23 (in Initial Reports of the Deep Sea Drilling Project). Washington, D.C., U.S. Govt. Printing Office, v. 23, p. 939-942.

Lloyd, R. M., 1973: Interstitial water studies, leg 15, δO^{18} in sulfate ion (in Initial Reports of the Deep Sea Drilling Project). Washington, D.C., U.S. Govt. Printing Office, v. 20, p. 887-889.

Lyon, G. L., 1973: Interstitial water studies, leg 15, chemical and isotopic composition of gases from Cariaco trench sediments (in Initial Reports of the Deep Sea Drilling Project). Washington, D.C., U.S. Govt. Printing Office, v. 20, p. 773-774.

F. PRECIPITATION

General

Bleeker, W., Dansgaard, W., and Lablans, W. N., 1966: Some remarks on simultaneous measurements of particulate contaminants including radioactivity and isotopic composition of precipitation. Tellus, v. 18, n. 4, p. 773-785.

Botter, R., Lorius, C., and Nief, G., 1961: Sur la teneur en deutérium des précipitation en terre de Victoria, Adélie, Antarctique (On the deuterium content of precipitations in Victoria, Adelie, Antarctic). Paris, Acad. Sci. Comptes Rendus, v. 251, n. 4, p. 573-575.

Brinkmann, R., Eichler, R., Ehhalt, D., and Münnich, K. O., 1963: The content of deuterium in rain and ground water. Naturwissenschaften, v. 50, p. 611-612.

Christiansen, W. N., Crabtree, R. W., and Laby, T. H., 1935: Density of light water; ratio of deuterium to hydrogen in rain water. Nature, v. 135, n. 3421, p. 870.

Cortecci, G., and Longinelli, A., 1970: Isotopic composition of sulfate in rain water, Pisa, Italy. Earth Planet. Sci. Lett., v. 8, n. 1, p. 36-40.

Dansgaard, W., 1964: Stable isotopes in precipitation. Tellus, v. 16, n. 4, p. 436-468.

Ehhalt, D., Knott, K., Nagel, J. F., et al., 1963: Deuterium and oxygen-18 in rain water. J. Geophys. Res., v. 68, n. 13, p. 3775-3780.

South Africa and Heidelburg, Germany

Ehhalt, D. H., and Ostlund, H. G., 1970: Deuterium in Hurricane Faith 1966--preliminary result. J. Geophys. Res., v. 75, n. 12, p. 2323-2327.

Friedman, I., Machta, L., and Soller, R., 1962: Water vapor exchange between a water droplet and its environment. J. Geophys. Res., v. 67, n. 7, p. 2761-2766.

Gambell, A. W., and Friedman, I., 1965: Note on the areal variation of deuterium/hydrogen ratios in rainfall for a single storm event. J. Appl. Meteorol., v. 4, n. 4, p. 533-535.

Eastern United States, November 1963

Gordiyenko, F. G., and Barkov, N. I., 1973: Variations of O^{18} content in the present precipitation of Antarctica. Sov. Antarct. Exped., Inform. Bull., v. 8, n. 9, p. 495-496.

International Atomic Energy Agency, 1970: Environmental isotope data no. 2, world survey of isotope concentration in precipitation (1964-1965). Vienna, IAEA, Tech. Rpt. Ser. 117.

International Atomic Energy Agency, 1971: Environmental isotope data no. 3, world survey of isotope concentration in precipitation (1966-1967). Vienna, IAEA, Tech. Rpt. Ser. 129, 402 p.

International Atomic Energy Agency, 1973: Environmental isotope data no. 4, world survey of isotope concentration in precipitation (1968-1969). Vienna, IAEA, Tech. Rpt. Ser. 147, 334 p.

Lorius, C., Merlivat, L., and Hagemann, R., 1969: Variation in the mean deuterium content of precipitations in Antarctica. J. Geophys. Res., v. 74, n. 28, p. 7027-7031.

Lorius, C., Merlivat, L., and Khagemann, R., 1970: Kolebaniya srednego soderzhaniya deyteriya v osadkakh v Antarktide (Oscillation in average deuterium content in Antarctic sands). Sov. Antarkt. Eksped. Inform. Byull., n. 77, p. 57-42.

Matsuo, S., and Friedman, I., 1967: Deuterium content in fractionally collected rain water. J. Geophys. Res., v. 72, n. 24, p. 6374-6376.

Merlivat, L., and Nief, G., 1967: Fractionnement isotopique lors des changements d'état solide-vapeur et liquide-vapeur de l'eau à des témpératures inférieures à 0°C (Isotopic fractionation at the time of changes of solid-vapor and liquid-vapor state of water at temperatures lower than 0°C). Tellus, v. 19, n. 1, p. 122-127.

Miyake, Y., Matsubaya, O., and Nishihara, C., 1968: Isotopic study on meteoric precipitations. Tokyo, Paper Meteor. Geophys., v. 19, n. 2, p. 243-266.

Mizutani, Y., and Rafter, T. A., 1969: Oxygen isotope composition of sulphates; pt. 5, isotopic composition of sulphate in rain water, Gracefield, New Zealand. N.Z. J. Sci., v. 12, n. 1, p. 69-80.

Moser, H., Silva, C., Stichler, W., and Stowhas, 1972: Measuring the isotope content in precipitation in the Andes (in The Role of Snow and Ice in Hydroldings of Banff Symposium, September 1972). Internat. Assoc. Hydrol. Sci. Pub. 107, v. 1, p. 14-23.

Rosinski, J., Langer, G., and Bleck, R., 1969: Removal of aerosol particles and fractional separation of HDO-H O during snowstorms. J. Atmos. Sci., v. 26, n. 2, p. 289-301.

Colorado

Tsunogai, S., and Matsubaya, O., 1967: Comments on K. Isono, M. Komabayasi and T. Takahashi--physical study of solid precipitation from convective clouds over the sea; pt. 1, deuterium content of snow crystals with respect to crystal slope and their relation to origins of the water vapour of snowfall; with reply by Isono, Komabayosi and Takahashi. Tokyo, Meteorol. Soc. Japan, Ser. 2, v. 45, n. 2, p. 200-204.

Woodcock, A. H., and Friedman, I., 1963: The deuterium content of raindrops. J. Geophys. Res., v. 68, n. 15, p. 4477-4483.

Hawaii

Hailstones

Bailey, I. H., and Macklin, W. C., 1965: Comments on L. Facy et al., study of the formation of hailstones by means of isotopic analysis; with reply by Facy et al. J. Geophys. Res., v. 70, n. 2, p. 493-499.

Bailey, I. H., et al., 1969: On the isotopic composition of hailstones. J. Atmos. Sci., v. 26, n. 4, p. 689-694.

Facy, L., 1962: Conditions de formation de la grêle déduites du rapport deutérium/hydrogène (Conditions for formation of hail deduced from the deuterium/hydrogen ratio). Genoa, Geofisica e Meteorlogia, v. 10, n. 1/2, p. 1-5.

Facy, L., Merlivat, L., Nief, G., et al., 1962: Etude de la formation d'un grêlon par une méthode d'analyse isotopique (Study of the formation of a hailstone with isotopic analysis method). J. Mécanique et Physique de l'Atmosphère, v. 4, n. 14, p. 67-78.

Facy, I., Merlivat, L., Nief, G., et al., 1963: Study of the formation of a hailstone by means of isotopic analysis. J. Geophys. Res., v. 68, n. 13, p. 3841-3848.

Jouzel, J., Merlivat, L., and Admirat, P., 1973: Etude isotopique de l'orage à grêle du 19 août dans la région de Clermont-Ferrant (Isotopic study of the hailstorm of August, 1971 in the Clermont-Ferrand region. J. Recherches Atmosphériques, v. 7, n. 3, p. 167-174.

Macklin, W. C., Merlivat, L., and Stevenson, C. M., 1970: Analysis of a hailstone. Royal Meteorol. Soc., Quart. Jour., v. 96, n. 409, p. 472-486.

Wales

Majzoub, M., Nief, G., and Roth, E., 1969: Variations and comparisons of deuterium and oxygen-18 concentrations in hailstones (<u>in</u> International Conference on Cloud Physics, Toronto, August 1968). Toronto, p. 450-454.

Pontikis, C., and Rigaud, A., 1972: Sur une méthode d'analyse isotopique des grêlons (Methods of isotopic analyses of hailstones). Paris, La Météorologie, Sér. 5, n. 23, p. 51-56.

G. GLACIERS AND SNOWPACKS

Aldaz, L., and Deutsch, S., 1967: On a relationship between air temperature and oxygen isotope ratio of snow and firn in the South Pole region. Earth Planet. Sci. Lett., v. 3, n. 3, p. 267-274.

Ambach, W., Dansgaard, W., Eisner, H., and Moller, J., 1968: The altitude effect on the isotopic composition of precipitation and glacier ice in the Alps. Tellus, v. 20, n. 4, p. 595-600.

Ambach, W., Eisner, H., Moser, H., and Stichler, W., 1970: Deuteriumgehalt des wassers im gletscherabfluss (Deuterium content of the water in glacier outflow). Naturwissenschaften, v. 57, n. 2, p. 86.

Ambach, W., and Dansgaard, W., 1970: Fallout and climate studies on firn cores from Carrefour, Greenland. Earth Planet. Sci. Lett., v. 8, n. 4, p. 311-316.

Ambach, W., Eisner, H., and Pessl, K., 1973: Isotopic composition of firn, old snow and precipitation in the Alpine region. Z. Gletscherk. Glazialgeol., v. 8, n. 1-2, p. 125-135.

Ambe, M., 1966: Deuterium content of water substances in Antarctica; pt. 1, geochemistry of deuterium in natural water on the east Ongul island. Japan Antarctic Res. Exped. (1956-62). Sci. Rep., Ser. C, n. 6, 13 p.

Arnason, B., 1969: The exchange of hydrogen isotopes between ice and water in temperature glaciers. Earth Planet. Sci. Lett., v. 6, n. 6, p. 423-430.

Iceland

Arnason, B., 1970: Exchange of deuterium between ice and water in glaciological studies in Iceland (in Isotope Hydrology, Proceedings, Symposium of International Atomic Energy Agency and UNESCO, March 9-13, 1970). Vienna, IAEA, STI/PUB/255, Paper n. SM-129/5, p. 57-71.

Arnason, B., Buason, T., Martinec, J., and Theodorsson, P., 1973: Movement of water through snowpack traced by deuterium and tritium (in The Role of Snow and Ice in Hydrology, Proceedings of Banff symposium, September 1972). Internat. Assoc. Hydrol. Sci. Pub. 107, v. 1, p. 299-312.

Ault, W. W., 1957: Oxygen isotope measurements on Arctic cores (in Age Measurements and Other Isotopic Studies on Arctic Materials). Columbia Univ., Lamont Geophysical Observatory, Contract AF-19(604)-1064, Final Rpt., 45 p.

Botler, R., Lorius, C., and Nief, G., 1961: Sur la datation des couches de névé dans l'Antarctique (On the dating of névé layers in the Antarctic). Paris, Acad. Sic. Comptes Rendus, v. 252, n. 3, p. 437-439.

Bray, J. R., 1970: Solar activity index; validity supported by oxygen isotope dating. Science, v. 168, n. 3931, p. 571-572.

Greenland

Craig, H., 1966: Discussion of paper by S. Epstein, R. P. Sharp and A. J. Gow, "Six-year record of oxygen and hydrogen isotope variation in South Pole firn." J. Geophys. Res., v. 71, n. 4, p. 1287-1288.

Dansgaard, W., Nief, G., and Roth, E., 1960: Isotope distribution in a Greenland iceberg. Nature, v. 185, n. 4708, p. 232.

Dansgaard, W., 1961: The isotopic composition of natural waters with special reference to the Greenland ice cap. Russian and Danish summaries. Medd. om Grønland, v. 165, n. 2, 120 p.

Dansgaard, W., 1969: Oxygen-18 analysis of water (in Etudes physiques et Chimiques sur la Glace de l'Indlandsis du Groenland 1959). Medd. Grönland, v. 177, n. 2, p. 33-36.

Dansgaard, W., Merlivat, L., and Roth, E., 1969: Stable isotope studies on EGIG samples, deuterium and O^{18} (in Etudes Physiques et Chimiques sur la Glace de l'Indlandsis du Groenland 1959). Medd. Grönland, v. 177, n. 2, p. 62-76.

Deutsch, S., Ambach, W., and Eisner, H., 1966: Oxygen isotope study of snow and firn on an Alpine glacier. Earth Planet. Sci. Lett., v. 1, n. 4, p. 197-201.

Epstein, S., and Sharp, R. P., 1959: Oxygen isotope variations in the Malaspina and Saskatchewan glaciers. J. Geol., v. 67, n. 1, p. 88-102.

Alaska and Alberta

Epstein, S., 1961: Oxygen isotope measurements in glacial ice. Washington, D.C., Natl. Res. Coun., Pub. 845, p. 102-105.

Greenland

Epstein, S., et al., 1963: Oxygen isotope ratios in Antarctic snow, firn and ice. J. Geol., v. 71, n. 6, p. 698-720.

Epstein, S., Sharp, R. P., and Gow, A. J., 1965: Six-year record of oxygen and hydrogen isotope variations in South Pole firn. J. Geophys. Res., v. 70, n. 8, p. 1809-1814.

Epstein, S., and Sharp, R. P., 1967: Oxygen and hydrogen isotope variations in a firn core, Eights Station, western Antarctica. J. Geophys. Res., v. 72, n. 22, p. 5595-5598.

Friedman, I., Schoen, B., and Harris, J., 1961: The deuterium concentration in Arctic Sea ice. J. Geophys. Res., v. 66, n. 6, p. 1861-1864.

Friedman, I., and Smith, G. I., 1970: Deuterium content of snow cores from Sierra Nevada area. Science, v. 169, n. 3944, p. 467-470.

Friedman, I., and Smith, G. I., 1972: Deuterium content of snow as an index to winter climate in the Sierra Nevada area. Science, v. 176, n. 4036, p. 790-793.

Gonfiantini, R., and Picciotto, E. E., 1959: Oxygen isotope variations in Antarctic snow samples. Nature, v. 184, n. 4678, Suppl. 20, p. 1557-1558.

Gonfiantini, R., et al., 1963: Snow stratigraphy and oxygen isotope variations in the glaciological pit of King Baudouin station, Queen Maud land, Antarctica. J. Geophys. Res., v. 68, n. 13, p. 3791-3798.

Gonfiantini, R., Togliatti, V., Tongiorgi, E., deBreuck, W., and Picciotto, E. E., 1963: Geographic variations of oxygen-18/oxygen-16 ratio in surface snow and ice from Queen Maud land, Antarctica. Nature, v. 197, n. 4872, p. 1096-1098.

Gonfiantini, R., 1965: Some results on oxygen isotope stratigraphy in the deep drilling at King Baudouin station, Antarctica. J. Geophys. Res., v. 70, n. 8, p. 1815-1819.

Judy, C., and Meiman, J. R., 1970: Deuterium variations in an annual snowpack. Water Resources Res., v. 6, n. 1, p. 125-129.

Colorado

International Geophysical Year Bulletin (no. 21), 1959: Oxygen isotope studies. Am. Geophys. Union, Trans., v. 40, n. 1, p. 81-84.

Antarctica, Greenland, Washington

Koerner, R. M., Paterson, W. S. B., and Krouse, H. R., 1973: δO^{18} profile in ice formed between the equilibrium and firn lines. Nature, Phy. Sci., v. 245, n. 148, p. 137-140.

Northwest Territories

Krouse, H. R., and Smith, J. L., 1973: O-18/O-16 abundance variations in Sierra Nevada seasonal snowpacks and their use in hydrological research (in The Role of Snow and Ice in Hydrology, Proceedings of Banff Symposium, September 1972). Internat. Assoc. Hydrol. Sci. Pub. 107, v. 1, p. 24-38.

Sierra Nevada, California

Lorius, C., 1961: Concentration en deutérium des couches de névé dans l'Antarctique (Concentration of deuterium in layers of névé in the Antarctic). Annales Géophysique, v. 17, n. 4, p. 378-387.

Lorius, C., and Merlivat, L., 1963: Sur la caractérization de la glace des inlandsis à partir de la teneur en deutérium (On the characterization of the ice of the ice caps on the basis of the deuterium content). Paris, Acad. Sci. Comptes Rendus, v. 256, n. 2, p. 475-477.

Lorius, C., 1964: Isotope in relation to polar glaciology. Polar. Rec., v. 12, n. 77, p. 211-228 (SCAR Bull., n. 17).

Macpherson, D. S., and Krouse, H. R., 1967: O^{18}/O^{16} ratios in snow and ice of the Hubbard and Kaskawulsh glaciers. Am. Geophys. Union, Geophys. Mon. Ser., n. 11, p. 180-194.

Meiman, J., Friedman, I., and Hardcastle, K., 1973: Deuterium as a tracer in snow hydrology (in The Role of Snow and Ice in Hydrology, Proceedings of Banff Symposium September 1972). Internat. Assoc. Hydrol. Sci. Pub. 107, v. 1, p. 39-50.

Colorado

Meiman, J. R., Friedman, I., and Hardcastle, K., 1974: Deuterium in Rocky Mountain snowpacks (in Advanced Concepts and Techniques in the Study of Snow and Ice Resources, An Interdisciplinary Symposium, Monterey, California, December 2-6, 1973). Washington, D.C., Natl. Acad. Sci., p. 152-160.

Merlivat, L., Lorius, C., Majzoub, M., Nief, G., and Roth, E., 1967: Isotopic examination at depth of an Antarctica glacier (in Proceedings of Symposium on Isotopes in Hydrology, International Atomic Energy and IUGG, Vienna, November 14-18, 1966). Vienna, IAEA, p. 671-679.

Merlivat, L., Lorius, C., and Nief, G., 1966: Etude isotopique d'un glacier en Antarctique (Isotope study of a glacier in the Antarctic). Acad. Sci. Comptes Rendus, Sér. B, v. 263, p. 414-416.

Merlivat, L., Ravoire, J., Vergnaud, J. P., and Lorius, C., 1973: Tritium and deuterium content of the snow in Greenland. Earth Planet. Sci. Lett., v. 19, n. 2, p. 235-240.

Merlivat, L., and Lorius, C., 1973: Teneur en deutérium et tritium de la neige du Groenland (The deuterium and tritium contents of snow in Greenland), abs. Réun. Annu. Sci. Terre, Programme Résumés, p. 299.

Moser, H., and Stichler, W., 1970: Deuterium measurements on snow samples from the Alps (in Isotope Hydrology Proc., Symposium of International Atomic Energy Agency and UNESCO, Vienna, March 9-13, 1970). Vienna, IAEA, STI/PUB/255, Paper n. SM-129/4, p. 43-57.

Moser, H., Rauert, W., Stichler, W., et al., 1973: Messungen des deuterium- und tritiumgehaltes von scheelsund schmelzwasserproben des Hintereisferners, Otztaler Alps (Measurements of deuterium and tritium contents of snow, ice and meltwater from Hintereis glacier, Otztal Alps). English and French summaries. Z. Gletscherk. Glazialgeol., v. 8, n. 1-2, p. 275-281.

Nief, G., 1969: Détérmination de la concentration en deutérium des échantillons (in Etudes Grönland), v. 177, n. 2, p. 36-42.

Oeschger, H., and Roth, E., 1969: Schlussbemerkungen, concluding remarks (in Études Physiques et Chimiques sur la Glace de l'Indlandsis du Groenland). Medd. Grönland, v. 177, n. 2, p. 111-113.

Oeschger, H., Stauffer, B., and Langway, C. C., Jr., 1970: Carbon dating of ice and other isotope studies at Byrd Station, Antarctica. Antarctic J., v. 5, n. 4, p. 112.

Picciotto, E. E., Maere, X. de, and Friedman, I., 1960: Isotopic composition and temperature of formation of Antarctic snows. Nature, v. 187, n. 4740, p. 857-859.

Picciotto, E., 1962: Notes on isotope glaciology. Cambridge Polar Record, v. 11, n. 71, p. 206-208.

Picciotto, E., et al., 1968: Determination of the rate of snow accumulation at the pole of relative inaccessibility, eastern Antarctica, a comparison of glaciological and isotopic methods. J. Glaciology, v. 7, n. 50, p. 273-287.

Renaud, A., 1969: Observations sur la glace surimposée (in Etudes Physiques et Chimiques sur la Glace de l'Indlandsis du Groenland 1959). Medd. Grönland, v. 177, n. 2, p. 93-100.

Renaud, A., 1969: Observations sur la glace bordière (in Etudes Physiques et Chimiques sur la Glace de l'Indlandsis du Groenland 1959). Medd. Grönland, v. 177, n. 2, p. 107-110.

Robin, G. de Q., 1972: Polar ice sheets, a review. Polar Record, v. 16, n. 100, p. 5-22.

Sharp. R. P., and Epstein, S., 1958: Oxygen isotope ratios and glacier movement (in Symposium of Chamonix, Physics of the Movement of the Ice, 1958). Internat. Assoc. Sci. Hydrol. Pub. 47, p. 359-369.

Sharp, R. P., et al., 1960: Oxygen isotope ratios in the blue glacier, Olympic Mountains, Washington, USA. J. Geophys. Res., v. 65, n. 12, p. 4043-4059.

Sharp, R. P., and Epstein, S., 1962: Comments on annual rates of accumulation in west Antarctica. Internat. Assoc. Sci. Hydrol., Pub. 58, p. 273-285.

Theordorsson, P., 1971: Rannsóknir á Bárdarbungu 1969 og 1970 (Investigations on Bardarbungu in 1969 and 1970), English summary. Jökull, v. 20, p. 1-14.

Thompson, L. G., Hamilton, W. L., and Bull, C., 1973: Analysis of the concentration of microparticles in the long ice core from Byrd Station. Antarctic J., v. 8, n. 6, p. 340-341.

Vergnaud, J. P., Botter, R., Lorius, C., et al., 1973: Tritium and deuterium content of snow in Greenland (in Symposium on Hydrogeochemistry and Biogeochemistry), v. 1, p. 621-632.

Vilensky, V. D., Teys, R. V., Yemel'yanov, V. V., and Kochetkova, S. N., 1972: Ispol'zovaniye izotopnykh metodov dlya opredeleniya sovremennoy skorsti nakopleniya snega v antarktide (Use of isotopic methods to determine present rates of snow accumulation in Antarctica). Geokhim., n. 9, p. 1071-1082.

Vilensky, V. D., Teys, R. V., and Kochetkova, S. N., 1974: Izotopnyy sostav kisloroda v snezhnom pokrove nekotorykk rayonov vostochnoy Antarktidy (Isotope composition of oxygen in snow of some regions of eastern Antarctica), English summary. Geokhim. Acad. Nauk SSSR, n. 1, p. 39-44.

West, K. E., and Krouse, H. R., 1972: Abundances of isotopic species of water in the St. Elias Mountains (in Icefield Ranges Research Project Scientific Results). New York, Am. Geographical Soc., Montreal, Canada, Arctic Inst. N. Am., p. 117-130.

Yukon Territory

Evaporation and Transpiration

Atomic Energy of Canada, Ltd., 1973: Biology and health physics division progress report, January 1 to March 31, 1973. Chalk River, Canada, Atomic Energy of Canada, Ltd. Nuclear Labs. (NTIS), 61 p.

Perch Lake, Ontario

Craig, H., Gordon, L. I., and Horibe, Y., 1963: Isotopic exchange effects in the evaporation of water. J. Geophys. Res., v. 68, n. 17, p. 5079-5087.

Ehhalt, D., and Knott, K., 1965: Kinetische isotopentiennung bei der verdampfung von wasser (Kinetic separation of isotopes during the evaporation of water). Tellus, v. 17, p. 389-397.

Fontes, J.-C., and Gonfiantini, R., 1967: Comportement isotopique au cours de l'évaporation de deux bassins sahariens (Isotopic behavior during evaporation of two Sahara basins). Earth Planet. Sci. Lett., v. 3, n. 3, p. 258-266.

Lloyd, R. M., 1966: Oxygen isotope enrichment of sea water by evaporation. Geochim. Cosmochim. Acta, v. 30, n. 8, p. 801-814.

Merlivat, L., and Coantic, M., 1975: Study of mass transfer at the air-water interface by an isotopic method. J. Geophys. Res. (in press).

Zimmerman, U., Ehhalt, D., and Münnich, K., 1967: Soil-water movement and evapotranspiration; changes in the isotopic composition of the water (in Isotopes in Hydrology, Proceedings Symposium of International Atomic Energy Agency and International Union Geodesy and Geophysics, Vienna, November 14-18, 1966). Vienna, IAEA, STI/PUB/141, Paper n. SM 83/38, p. 567-585.

Central Europe

I. LAKES AND RESERVOIRS

Bonner, F. T., et al., Chlorine-36 and deuterium study of Great Basin Lake waters. Geochim. Cosmochim. Acta, v. 25, n. 4, p. 261-266.

California, Utah and Nevada

Cortecci, G., 1973: Oxygen isotope variations in sulfate ions in the water of some Italian lakes. Geochim. Cosmochim. Acta, v. 37, n. 6, p. 1531-1542.

Dincer, T., and Payne, B. R., 1967: An isotope survey of lakes in the karst region of southern Turkey (in Hydrology of Fractured Rocks), French summary. Internat. Assoc. Sci. Hydrol., Pub. No. 74, p. 654-661.

Dincer, T., 1968: The use of oxygen-18 and deuterium concentrations in the water balance of lakes. Water Resources Res., v. 4, n. 6, p. 1289-1306.

Fontes, J. Ch., Gonfiantini, R., and Roche, M. A., 1970: Deuterium and oxygen-18 in water of Lake Chad (in Isotope Hydrology, Proceedings Symposium of International Atomic Energy Agency and UNESCO, Vienna, March 9-13, 1970). Vienna, IAEA, STI/PUB/255, Paper n. SM-129/23 (in French), p. 387-404.

Friedman, I., Norton, D., Carter, D. B., and Redfield, A. C., 1956: The deuterium balance of Lake Maracaibo. Limnology and Oceanography, v. 1, n. 4.

Caribbean Sea

Friedman, I., and Redfield, A. C., 1967: Application of deuterium analyses to the hydrology of the lakes of the Grand Coulee, Washington (in Isotope Techniques in the Hydrologic Cycle). Am. Geophys. Union, Geophys. Mon. Ser., n. 11, p. 77-80.

Friedman, I., and Redfield, A. C., 1971: A model of the hydrology of the lakes of the lower Grand Coulee, Washington. Water Resources Res., v. 7, n. 4.

Gat, J. R., 1970: Environmental isotope balance of Lake Tiberias (in Isotope Hydrology, Proceedings Symposium of International Atomic Energy Agency and UNESCO, Vienna, March 9-13, 1970). Vienna, IAEA, STI/PUB/255, Paper n. SM-129/8, p. 109-127.

Merlivat, L., 1970: Quantitative aspects of the study of water balances in lakes using the deuterium and oxygen-18 concentrations in the water (in Isotope Hydrology, Proceedings Symposium of International Atomic Energy Agency and UNESCO, Vienna, March 9-13, 1970). Vienna, IAEA, STI/PUB/255, Paper n. SM-129/7, p. 87-107.

Merlivat, L., 1970: Problems associated with the quantitative study of lake balance by means of the concentrations of deuterium and /sup 18/0 in the water. NTIS, CEA-CONF-1538, 29 p.

Payne, B. R., 1970: Water balance of Lake Chala and its relation to ground water from tritium and stable isotope data. J. Hydrol., v. 11, n. 1, p. 47-58.

Ragotzkie, R. A., and Friedman, I., 1965: Low deuterium content of Lake Vanda, Antarctica. Science, v. 148, n. 3674, p. 1226-1227.

Soyfer, V. N., Brezgunov, V. S., Verbolob, K., Votintsev, K., and Romanov, V. V., 1970: Application of the isotopic method to study of mixing processes in Lake Baikal (Primeneniye izotopnogo metoda dlya izucheniya protsessov vodoobmena ozera baykal) (in Techeniya i diffuziya vod Baykala). Akad. Nauk SSSR Sibirskoyo Otdeleniye, Limnologicheskiy Institut Trudy, v. 14, n. 34, p. 146-153.

Takahashi, T., Broecker, W., and Hui Li, Y., 1968: Chemical and isotopic balances for a meromictic lake. Limnol. and Oceanogr., v. 13, n. 2, p. 272-292.

Green Lake, New York

Zimmerman, U., and Ehhalt, D. H., 1970: Stable isotopes in study of water balance of Lake Neusiedl, Austria; investigation of the reliability of the stable isotope method (in Isotope Hydrology, Proceedings Symposium of International Atomic Energy Agency and UNESCO, Vienna, March 9-13, 1970). Vienna, IAEA, STI/PUB/255, Paper n. SM-129/9, p. 129-138.

J. RIVERS

Eichler, R., 1967: Deuterium-gehalt in alpenwässern (Deuterium content in alpine waters), English, French and Russian summaries. Geol. Rundsch., v. 56, n. 3, p. 876-881.

Matsui, E., Salati, E., Brinkmann, W. L. F., and Friedman, I., 1972: Relative rates of flow on the Negro and Solimoes Rivers determined through O^{18} concentrations (Portuguese). Acta Amazonica, v. 2, n. 3, p. 31-46.

K. UNSATURATED ZONE

Stewart, G. L., 1967: Fractionation of tritium and deuterium in soil water (in Isotope Techniques in the Hydrologic Cycle). Am. Geophys. Union, Geophys. Mon. Ser., n. 11, p. 159-168.

Zimmerman, U., Münnich, K. O., and Roether, W., 1967: Downward movement of soil moisture traced by means of hydrogen isotopes (in Isotope Techniques in the Hydrologic Cycle). Am. Geophys. Union, Geophys. Mon. Ser., n. 11, p. 28-36.

L. GROUND WATER

General

Batsche, H., Stichler, W., and Moser, H., 1972: Messungen des deuterium-und sauerstoff-18-gehalts in karstwaessern (Measurements of deuterium and oxygen-18 contents of karst waters), with discussion, (*in* Internationale Fachtagung zur Untersuchung Unterirdischer Wasserwege Mittels Kuensthicher und Natuerlicher Markierungsmittel, 2nd, Vortraege diskussionen und Beitraege), English, French and Russian summaries. Ger., Bundesanst. Bodenforsch. Geol. Landesaemt., Geol. Jahrb., Reihe C, n. 2, p. 275-288.

Gat, J. R., 1971: Comments on the stable isotope method in regional ground water investigations. Water Resources Res., v. 7, n. 4, p. 980-993.

Mook, W. G., 1972: Application of natural isotopes in ground water hydrology. Netherlands, Geologie en Mijnbouw, v. 51, n. 1, p. 131-136.

Moser, H., and Neumaier, F., 1973: Grundwasser; untersuchung mit isotopen (Ground water isotopic studies), English summary. Unschau, v. 73, n. 16, p. 501-502.

Moser, H., Neumaier, F., and Rauert, W., 1973: Anwendung von isotopenmethoden zur untersuchung der verunreinigung von grundwasser (Applications of isotope methods for the investigation of ground water pollution), English summary. Dtsch. Geol. Ges., Z., v. 124, pt. 2, p. 501-514.

Vogel, J. C., and Ehhalt, D., 1963: The use of carbon isotopes in ground water studies (*in* Radioisotopes in Hydrology). Vienna, IAEA, p. 387-395.

Areal Studies

Alekseyev, F. A., Gottikh, R. P., and Taneyen, R. N., 1970: Podzemnyye vody Bukharo-Karshinskogo artezianskogo basseyna v svete izucheniya radioaktiunykh elementov i deyteriya (Ground water in the Bukhara-Karshi artesian basin in view of the study of radioactive elements and deuterium). Sov. Geol., n. 10, p. 44-58.

Alekseyev, F. A., Gottikh, R. P., Soyner, V. N., and Brezgunov, V. S., 1970: Radioaktivnyye elementy i deyteriy v podzemnykh vodakh bukharokarshinskogo artezianskogo basseyna (Radioactive elements and deuterium in ground waters of the Bukharokarshinskiy artesian basin). Geokhimiya, n. 12, p. 1483-1494.

Bedmar, A. P., Sanchez, W., Castagnet, A. C., et al., 1973: Estudo dos aquiferos das bacias dos Rios Gurgueia e Fidalgo, Piaui utilization isotopos ambientais (The study of the aquifers of the basins of Gurgueia River and Fidalgo River, Piaui, Brazil, using ambient isotopes) (in Resumo das Comunicacoes Sessoes Technicas). Geofisica, Congr. Bras. Geol., n. 27, Bol. 1, p. 189.

Blavouv, B., Fontes, J. C., Jacob, P., et al., 1973: Récherches hydrochimiques et isotopiques sur les systèmes aquifères du massif du Mont Dore (Hydrochemical and isotopic studies on the aquifer systems of Mont-Dore Massif, France), abs. Réun. Annu. Sci. Terre, Programme Résumés, p. 83.

Bortolami, G., Fontes, J. C., and Panichi, C., 1973: Isotopes du milieu et circulations dans les aquifères du sous-sol venitien (Isotopic study and circulation of aquifers in the subsoil of Venice), English summary. Earth Planet. Sci. Lett., v. 19, n. 2, p. 154-167.

Conrad, G., Fontes, J.-C., Létolle, R., and Roche, M., 1966: Etude isotopique de l'oxygène des eaux de la région de la Haute Saoura, Sahara nord-occidental (Oxygen isotope study of the waters of the Haute Saoura, northwest Sahara). Acad. Sci. Comptes Rendus, Sér. D, v. 262, n. 10, p. 1058-1061.

Cotecchia, V., and Tongiorgi, E., 1963: Natural tracing by seasonal variation of O-18/O-16 ratio in the ground water research. Italy, Com. Naz. Energia Nucleare Rap. Tec., Ser. 10, RT/Geo. 25, 9 p.

Solentine Peninsula, Italy

Davis, G. H., Meyer, G. L., and Yen, C. K., 1968: Isotope hydrology of the artesian aquifers of the Styrian basin, Austria, French and German summaries. Steirische Beitr. Hydrogeol., v. 20, p. 51-62.

Davis, G. H., Lee, C. K., Bradley, E., et al., 1970: Geohydrologic interpretations of a volcanic island from environmental isotopes. Water Resources Res., v. 6, n. 1, p. 99-109.

Cheju Island, Korea

Dincer, T., Payne, B. R., Yen, C. K., et al., 1972: Das tote gebirge als entwaesserungstypus der karstmassive der nordoestlichen kalkhochalpen, ergebnisse von isotopenmessungen (The Tote Gebirge as drainage model for the karst drainage of northeastern limestone Alps; isotope measurements), English summary. Steirische Beitr. Hydrogeol., v. 24, p. 71-109.

Dowgiallo, J., 1973: Wyniki badan skladu izotopowego tlenu i wodoru w wodach podziemnych Polski poludniowe (Isotopic composition of oxygen and hydrogen from ground water in southern Poland) (<u>in</u> Z. badan hydrogeologicznych w Polsce, tom III), English and Russian summaries. Pol., Inst. Geol., Biul., n. 277, p. 319-338.

Drost, W., 1970: Ground water measurements at the site of the Sylvenstein dam in the Bavarian Alps (<u>in</u> Isotope Hydrology, Proceedings Symposium of International Atomic Energy Agency and UNESCO, Vienna, March 9-13, 1970). Vienna, IAEA, STI/PUB/255, Paper n. SM-129/25, p. 421-437.

Fritz, P., 1968: Der isotopengehalt der mineralwasserquellen von Stuttgart und umgebung und ihrer mittelpleistozanen travertin-ablagerungen (The isotopic composition of mineral springs in the Stuttgart area and associated middle Pleistocene travertine deposits), English summary. Oberrhein. Geol. Ver., Jahresber. Mitt. v. 50, p. 53-69.

Hufen, T. H., Buddemeier, R. W., and Lau, L. S., 1974: Isotopic and chemical characteristics of high-level ground waters on Oahu, Hawaii. Water Resourses Res., v. 10, n. 2, p. 366.

Martinec, J., Siegenthaler, U., Oeschger, H., and Tongiorgi, E., 1974: New insights into the run-off mechanism by environmental isotopes (<u>in</u> Isotope Techniques in Ground Water Hydrology), Vienna, IAEA, v. 1.

Switzerland

Moser, H., and Stichler, W., 1972: Altergliederung von tiefliegerden artesischen wasser messung des deuterium-gehaltes (Age determination of deep-seated artesian waters by measurement of the deuterium content). Naturwissenschaften, v. 59, n. 3, p. 122-123.

Austria

Moser, H., and Stichler, W., 1975: Use of environmental isotope methods as a reconnaissance tool in ground water exploration near San Antonio de Pichincha, Ecuador, abs. EOS, v. 57, n. 5.

Moussu, H., 1973: Study of the sandy Maestrichian aquifer and Continental Terminal in Senegal with environmental isotopes and particularly with C^{14}, abs. (in IAEA Research Contracts, Thirteenth Annual Report). Radioisotopes and Radiation Applications, IAEA, Tech. Rpt. Ser. n. 144, p. 177-178.

Plata, A., Sanchez, W., and Szulak, C., 1973: Studies of aquifers of the Gargueia and Fidalgo River basins in the state of Piaui using environmental isotopes. Sao Paulo, Brazil, Instituto de Energia Atomica (NTIS), 46 p.

Senturk, F., Bursali, S., Omay, Y., Ertan, I., and Guler, S., 1970: Isotope techniques applied to ground water movement in the Konya plains (in Isotope Hydrology, Proceedings, Symposium of International Atomic Energy Agency and UNESCO, Vienna, March 9-13, 1970). Vienna, IAEA, STI/PUB/255, Paper n. SM-129/11, p. 153-161.

Turkey

Sultankhodzhayer, A. N., 1966: K kharakteristike izotopnogo sostave podzemnykh vod Ferganskogo artezianskogo basseyna (On the nature of the isotopic composition of ground waters of the Fergana basin). Uzbek. Geol. Zhur., n. 6, p. 67-72.

USSR

Sultankhodzhayer, A. N., Irmukhamedov, M. A., and Sabirov, K. A., 1972: Izotopnyy sostav kisloroda podzemnykk vod Ferganskogo artezianskogo basseyna (Isotopic composition of oxygen from ground water in the Fergana artesian basin). Uzbek summary. Uzbek. Geol. Zhur., n. 5, p. 29-32.

USSR

Surganova, N. A., Taneyev, R. N., Gottikh, R. P., et al., 1970: Zakonomernosti raspredeleniya radioaktivnykh elementov i deyteriya v vodakh Bukharo-Karshinskogo artezianskogo basseyna (Distribution of radioactive elements and deuterium in the waters of the Bukhara-Karshi artesian basin) (in Geokhimicheskiye Metody Poiskov Nefti i Gaza i Voprosy Yadernoy Geologii). Vses. Nauchno-Issled. Inst. Yad. Geofiz. Geokhim., Tr., n. 8, p. 324-332.

USSR

Vogel, J. C., Lerman, J. C., and Mook, W. G., 1972: Natural isotopes in the ground water of the Tulum Valley, San Juan, Argentina. Internat. Assoc. Hydrol. Sci., Assoc. Internat. Sci. Hydrol., Bull., v. 17, n. 1, p. 85-96.

Winograd, I. J., 1972: Deuterium as a tracer of regional ground water flow, southern Great Basin, Nevada and California. Geol. Soc. Am. Bull., v. 83, n. 12, p. 3691-3708.

Recharge

Conrad, G., 1973: La part des récharges paléoclimatiques dans l'alimentation des grandes nappes aquifères du Nord-Ouest du Sahara Algérian (The role of paleoclimatic changes in the recharge of major aquifers in the northwestern Algerian Sahara), abs. Soc. Géol. Fr. Bull., v. 15, suppl., p. 7.

Dincer, T., Al-Mugrin, A., and Zimmerman, U., 1974: Study of the infiltration with recharge through the sand dunes in arid zones with special reference to the stable isotopes and thermonuclear tritium. J. Hydrol., v. 23, p. 79-109.

Gat, J. R., and Tzur, Y., 1967: Modification of the isotopic composition of rainwater by processes which occur before ground water recharge (in Isotopes in Hydrology, Proceedings, Symposium of International Atomic Energy Agency and International Union Geophysics and Geodesy, Vienna, November 14-18, 1966). Vienna, IAEA, p. 49-60.

Israel

Margat, J., 1966: Age des eaux souterraines et renouvellement des réserves des nappes; réflexions sur les bases de l'hydrochronologie. Fr. Bur. Rech. Géol. Minières Bull., n. 6, p. 37-51.

Simpson, E. S., Thorud, D. B., and Friedman, I., 1970: Distinguishing seasonal recharge to ground water by deuterium analysis in southern Arizona (in World Water Balance - Bilan Hydrique Mondia), French, Spanish and and Russian summary). Internat. Assoc. Sci. Hydrol., UNESCO, Stud. Rep. Hydrol., n. 11, p. 623-633.

Simpson, E. S., Thorud, D. B., and Friedman, I., 1970: Distinguishing winter from summer recharge to ground water in southern Arizona by deuterium analysis (in Symposium on World Water Balance). Internat. Assoc. Sci. Hydrol. Pub. n. 92, p. 100-101.

Vogel, J. C., 1967: Investigation of ground water flow with radiocarbon (in Isotopes in Hydrology, Proceedings, Symposium of International Atomic Energy Agency and International Union Geophysics and Geodesy, Vienna, November 14-18, 1966). Vienna, IAEA, p. 355-368.

Netherlands

Brines

Brezgunov, V. S., Sultankhodzhayev, A. N., and Tyminskiy, V. G., 1967: Primeneniye yestestvennykh radioktivnykh i stabil'nykh izotopov dlya vyyasneniya usloviy formirovaniya mineral'nykh vod, na primere Pritashkentskogo artezianskogo basseyna (Use of natural radioactive and stable isotopes to bring out the conditions of formation of mineral waters, exemplified by the Tashkent artesian basin). Uzbek. Geol. Zhur., n. 5, p. 61-68.

Clayton, R., Friedman, I., Grof, D. L., Mayeda, T. K., Meents, W. F., and Shimp, N. F., 1966: The origin of saline formation water; 1. Isotopic composition. J. Geophys. Res., v. 71, n. 16, p. 3869-3882.

Illinois, Michigan, Gulf of Mexico Coast, Alberta

Dowgiallo, J., and Tongiorgi, E., 1972: The isotopic composition of oxygen and hydrogen in some brines from the Mesozoic in northwest Poland. Geothermics, v. 1, n. 2, p. 67-69.

Gat, J. R., Mazor, E., and Tzur, Y., 1969: The stable isotope composition of mineral waters in the Jordan Rift Valley in Israel. J. Hydrol., v. 7, n. 3, p. 334-352.

Graf, D. L., et al., 1965: The origin of saline formation waters; II. Isotopic fractionation by shale micropore systems. Illinois, State Geol. Survey, Circ. 393, 32 p.

Illinois, Michigan, Gulf of Mexico Coast, Alberta

Merlivat, L., and Vuillaume, Y., 1970: Caractérisation de l'intrusion marine dans la nappe de la Crau à l'aide du deutérium. France, Bull. B.R.G.M., Sec. 3, n. 2, p.81-85.

Pinneker, Ye. V., Soyfer, V. N., Brezgunov, V. S., and Vlasova, L. S., 1960: Ob izotopnom sostave vodoroda rassolov yuga Sibirskoy platformy (On the isotopic composition of hydrogen in brines from the southern part of the Siberian platform). Akad. Nauk SSSR Doklady, v. 179, n. 6, p. 1452-1455.

Polivanova, A. I., 1970: Isotopic composition of underground brines as a clue to the origin, abs. Geochem. Internat., v. 7, n. 4, p. 715-716.

USSR

Seletskiy, Yu. B., Polyakov, V. A., Yakubovskiy, A. V., et al., 1973: Genezis sverkhkrepkikh khloridno-kal'tsiyevykh rassolov Angaro-Lenskogo artesianskogo basseyna po dannym mass-spektrometricheskikh opredeleniy deyteriya i O^{18} (Genesis of supersaturated calcium chloride brine in the Angara-Lensk basin from the determination of deuterium of O^{18} with mass spectroscopy, abs). Mosk. Ovo. Ispyt. Prir., Byull., Otd. Geol., v. 48, n. 4, p.140.

Vetshteyn, V. Yu., Malyuk, G. A., and Lapshin, F. V., 1972: Izotopniy sklad kisnyu i vodnyu minniral'nikh vod Ukrains'kikh Karpat yak kriteriy ikh genezisu (Isotopic composition of oxygen and hydrogen from the mineral waters as a criterion of their genesis), English and Russian summary. Akad. Nauk Ukr. RSR, Dopov., Ser. B, n. 12, p. 1062-1066.

Associated with Oil, Gas and Other Hydrocarbons

Brezgunov, V. S., Vlasova, L. S., and Soyfer, V. N., 1968: Izotopnyy sostav vodoroda podzemnykh vod i neftey v svyazi s ikh proiskhozhdeniyem (Isotopic composition of hydrogen as a clue to the origin of ground water and petroleum). Geochem. Internat., v. 5, n. 1, p. 65-78.

Brezgunov, V. S., Vlasova, L. S., and Soyfer, V. N., 1968: Ixotopnyy sostav vodoroda podzemnykh vod i neftey v svyazi s ikh proiskhozhdeniyem (Isotopic composition of the hydrogen of ground waters and oils in relation to their origin). Geokhim., n. 1, p. 80-98.

Degans, E. T., Hunt, J. M., Reuter, J. H., and Reed, W. E., 1964: Data on the distribution of amino acids and oxygen-isotopes in petroleum brine waters of various geologic ages. Sedimentology, v. 3, n. 3, p. 199-225.

Gabril'yan, A. M., 1957: Nekotoryye dannyye o summarnom izotopnom sostave vod neftyanykh mestorozhdeniy Fergany (Some data on the total isotopic composition of the waters of the oil fields of Fargana). Akad. Nauk Uzbek. SSR Izv., Ser. Geol., n. 4, p. 47-55.

USSR

Kharaka, Yousif, K., Berry, F. A. F., and Friedman, I., 1973: Isotopic composition of oil field brines from Kettleman North Dome, California and their geologic implications. Geochim. Cosmochim. Acta, v. 37, n. 8, p. 1899-1908.

Kolodiy, V. V., and Dobrov, Yu. V., 1969: Deyteriy v podzemnykh vodakh Zapadno-Turkmenskoy neftegazonosnoy oblast.(Deuterium in ground waters of the West Turkmenian oil and gas district), English summary. Geokhim., n. 3, p. 341-347.

USSR

Malyuk, G. A., Trachuk, V. G., Vetshteyn, V. Yu., et al., 1973: Vmist deyteriyu ta kisnyu-18 u pidzemnikh rozsolakh naftovikh rodovishch Prip'yats'kogo proginu (Deuterium and O-18 in brines of the oil fields of the Pripet Basin, USSR), English and Russian summaries). Akad. Nauk Ukr. RSR, Dopov., Ser. B, n. 1, p. 26-29.

Subbota, M. I., 1968: Deyteriy v vodakh Dneprovsko-Donetskogo neftegazonogo basseyna i voprosy vodoobmena glubokikh gorizontov (Deuterium in waters of the Dneiper-Donets oil and gas bearing basin and some questions of deep water exchange). Sov. Geol., n. 11, p. 142-146.

USSR

Vetshteyn, V. Ye., Gutsalo, L. K., Malyuk, G. A., et al., 1973: K voprosu proiskhozhdeniya podzemnykh vod osadochnoy tolshchi Dneprovski-Donetskoy heftegazonosnoy provintsii po dannym izotopnogo sostava kisloroda i vodoroda (Genesis of ground waters in the sedimentary series of the Dnieper-Donets oil and gas bearing province, based on isotope composition data of oxygen and hydrogen). English summary. Geokhim., Akad. Nauk SSSR, n. 3, p. 327-338.

USSR

In Rock Reactions, In Magmas And In Juvenile Waters

Forester, R. W., and Taylor, H. P., Jr., 1972: Oxygen and hydrogen isotope data on the interaction of meteoric ground waters in a gabbro-diorite stock, San Juan Mountains, Colorado (in Geochemistry, Section 10). Internat. Geol. Congr., Proc., Congr. Geol. Internat. Programme, n. 24, p. 254-263.

Friedman, I., Lipman, P. W., Obradovich, J. D., et al., 1974: Meteoric water in magmas. Science, v. 184, n. 4141, p. 1069-1072.

Nevada, Colorado and Wyoming

Seletskiy, Yu. B., 1969: K voprosu o deyterii kak kriterii yuvenil'nosti i granitsakh primenimosti plotnostnykh metodov izucheniya izotopnogo sostave podzemnykh vod (On the problem of deuterium as a criterion of juvenility and the limits of application of density methods of studying the isotopic composition of ground waters). Moskov. Obshch. Ispytaleley Prirody Byull., Otdel Geol., v. 44, n. 3, p. 126-132.

Thermal Water

Arnason, B. : Measurements on the D/H ratio in hydrogen and water vapour collected at Surtsey. Surtsey Res. Soc., Prog. Rpt. 3, p. 93-94.

Arnason, B., 1964: Measurements on the D/H ratio in H - gas and water vapour collected at the volcanic island Surtsey during the year 1964. Surtsey Res. Soc., Prog. Rpt. 1, p. 27-33.

Arnason, B., 1967: Deuterium content of water vapour and hydrogen in volcanic gas at Surtsey, abs. Surtsey Res. Conf. Proc., p. 70-71.

Arnason, B., and Sigurgeirsson, T., 1968: Deuterium content of water vapour and hydrogen in volcanic gas at Surtsey Island. Geochim. Cosmochim. Acta, v. 32, n. 8, p. 807-813.

Arnason, B., 1969: Tvívetni í grunnvatni og jöklum á Islandi (Deuterium in ground water and glaciers in Iceland), English summary. Jökull, v. 18, p. 337-349.

Arnason, B., and Tomasson, J., 1973: Deuterium and chloride in geothermal studies in Iceland (in Proceedings of the United Nations Symposium on the Development and Utilization of Geothermal Resources, Pisa, Italy, September 22 to October 1, 1970). Geothermics, Spec. Issue 2, v. 2, pt. 2, p. 1405-1415.

Babinets, A. Ye., Lyal'ko, V. I., and Mitnik, M. M., 1968: Vozmozhynyye puti resheniya nekotorykh voprosov formirovaniya termal'nykh vod v artezianskikh basseynakh s ispol' zovaniyem termogidrodinamicheskikh raschetov i izotopnykh sootnosheniy (Possible ways of solving some problems of the formation of thermal waters in artesian basin using thermohydrodynamic calculations and isotopic ratios). (in Internat. Geol. Congr., 23rd, Prague, Proceedings, Symposium 2, Genesis of Mineral and Thermal Waters). Prague, Academia, p. 9-16.

Balke, K.-D., 1973: Geothermische und hydrogeologische untersuchungen in der suedlichen niederrheinischen bucht (Geothermal and hydrogeologic studies in the southern Lower Rhine Basin, Germany). Ger., Bundesanst. Bodenforsch. Geol. Landesaemt., Geol. Jahrd., Reihe C, n. 5, p. 5-61.

Banwell, C. J., 1967: Oxygen and hydrogen isotopes in New Zealand thermal areas (in Nuclear Geology on Geothermal Areas, Spoleto, 1963). Cons. Naz. Ric., Lab. Geol. Nucl., p. 95-138.

Baskov, Ye. A., Vetshteyn, V. Ye., Surikov, S. N., et al., 1973: Izotopnyy sostav H, O, C, Ar, He termal'nykh vod i gazov Kurilo-Kamchatskoy vulkanicheskoy oblasti kak pokazatel'usloviy ikh formirovaniya (Isotope composition of H, O, C, Ar, He in thermal waters and gases of the Kuril-Kamchatka volcanic region as an indicator of the conditions of their formation), English summary. Geokhim. Akad. Nauk SSSR, n. 2, p. 180-189.

Brezgunov, V. S., Lomonosov, I. S., Pinneker, Ye. V., et al., 1971: Izotopnyy sostav vodoroda termal'nykh i uglekislykh vod yuga vostochnoy Sibiri (Hydrogen isotope composition of hot springs and carbonic waters of south-eastern Siberia) (in Podzemnyye vody Sibiri i Dal'nego Vostoka). Akad. Nauk SSSR, Sib. Otdel. Kom. Izuch. Podzem. Sib. Dal'nego Vostoka, p. 89-93.

Boato, G., and Craig, H., 1955: Geochemistry of the isotopes in meteoric water and of thermal origin (in 1st Congr. Nuclear Geol., Rome, 1955). Comit. Naz. Ric. Nucleari, p. 22-29.

Bodvarsson, G., 1962: The use of isotopes of hydrogen and oxygen for hydrological purposes in Iceland. Jökull, v. 12, p. 49-52.

Cheminée, J., Létolle, R., and Olive, P., 1969: Premières données isotopiques sur des fumérolles de volcans italiens (First isotopic data on the fumaroles of Italian volcanoes). Bull. Volcanol., v. 32, n. 3, p. 469-475.

Coplen, T. B., 1973: Isotopic composition of calcite and water from the Dunes-DWR #1 geothermal test corehole, Imperial Valley, California, abs. EOS, Am. Geophys. Union, Trans., v. 54, n. 4, p. 488.

Cortecci, G., 1973: Oxygen-18 and carbon-13 contents of the sulfates and the carbonates associated with some oxidizing geothermal environment. Geothermics, v. 2, n. 2.

Craig, H., Boato, G., and White, D. E., 1954: Isotopic geochemistry of thermal waters (in Nuclear Processes in Geologic Settings). Natl. Acad. Sci., Res. Coun. Pub. 400, p. 29-38.

Craig, H., 1961: Isotopic geochemistry of volcanic water and steam, abs. Science, v. 134, n. 3488, p. 1427-1428.

Craig, H., 1966: Isotopic composition and origin of the Red Sea and Salton Sea geothermal brines. Science, v. 154, n. 3756, p. 1544-1548.

Craig, H., 1967: The isotopic geochemistry of water and carbon in geothermal (in Nuclear Geology on Geothermal Areas, Spoleto, 1957). Pisa, Italy, Laboratorio Geologia Nucleare, p. 17-53.

Craig, H., 1969: Discussion on source fluids for the Salton Sea geothermal system. Am. J. Sci., v. 267, p. 249-255.

Deuser, W. G., and Degens, E. T., 1969: O^{18}/O^{16} and C^{13}/C^{12} ratios of fossils from the hot brine deep area of the central Red Sea (in Hot Brines and Recent Heavy Metal Deposits in the Red Sea; A Geochemical and Geophysical Account). New York, Springer-Verlag, p. 336-347.

Ellis, A. J., 1971: Quantitative interpretation of chemical characteristics of hydrothermal systems (in United Nations Symposium of the Development and Utilization of Geothermal Resources, v. 2). Geothermics, Spec. Issue, n. 2, p. 516-528.

Ferrara, G. C., Ferrara, G., and Gonfiantini, R., 1963: Carbon isotopic composition of carbon dioxide and methane from steam jets of Tuscany (in Nuclear Geology on Geothermal Areas Symposium, Spoleto, Italy, September 9-13, 1963). Natl. Res. Coun., Nuclear Geology Lab. Pisa, p. 275-282.

Ferrara, G. C., Panichi, C., and Stefani, G., 1971: Remarks on the geothermal phenomenon in an intensively exploited field; results of an experimental well (in United Nations, Symposium on the Development and Utilization of Geothermal Resources, v. 2). Geothermics, Spec. Issue, n. 2, p. 578-586.

Italy

Florkowski, T., and Job, C., 1969: Origin and underground flow time of thermal waters in crystalline basement complexes. German summary. Steirische Beitr. Hydrogeol., v. 21, p. 37-50.

Fontes, J.-C., Glangeaud, L., Gonfiantini, R., and Tongiorgi, E., 1963: Composition isotopique et origine des eaux et gaz thermaux du Massif Central (Isotopic composition and origin of the thermal waters and gases of the Massif Central). Paris, Acad. Sci. Comptes Rendus, v. 256, n. 2, p. 472-474.

France

Friedman, I., 1970: Some investigations of the deposition of travertine from Hot Springs - I. The isotopic chemistry of the travertine-depositing spring. Geochim. Cosmochim. Acta, v. 34, p. 1303-1345.

Yellowstone National Park

Germanov, A. I., Mavritskiy, B. F., Pit'yeva, K. Ye., et al., 1972: Problema glubinnogo proiskhozdeniya termal'nykh podzemnykh vod v svete sovremennoy informatsii o verkhney gidrogeosfere (Deep source of thermal ground waters in the light of present-day data concerning the upper hydrosphere). Akad. Nauk SSSR, Izv., Ser. Geol., n. 8, p. 113-123.

Giggenbach, W., 1971: Isotopic composition of waters of the Broadlands geothermal fields. New Zealand J. Sci., v. 14, n. 4, p. 959-970.

Gonfiantini, R., Borsi, S., Ferrara, G., and Panichi, C., 1973: Isotopic composition of waters from the Danakil depression (Ethiopia). Earth Planet. Sci. Lett., v. 18, n. 1, p. 13-21.

Gunter, B. D., 1968: Geochemical and isotopic studies of hydrothermal gases and waters. Univ. Arkansas, Ph.D. Dissertation, 90 p.

Yellowstone National Park, Lassen Volcanic National Park

Gunter, B. D., and Musgrave, B. C., 1971: New evidence on the origin of methane in hydrothermal gases. Geochim. Cosmochim. Acta, v. 35, p. 113-118.

Halston, J. R., 1961: Isotope geology in the hydrothermal areas of New Zealand (<u>in</u> United Nations Conf. on New Sources of Energy). U.N., v. 2, p. 259.

Jacobshagen, V., and Münnich, K. O., 1964: Carbon-14 age determination and other isotope investigations on hot salt-water springs of the Ruhr carboniferous. Neues Jahrb. Geol., Palaont., Monatsh., n. 9, p. 566-568.

Job, C., and Zötl, J., 1969: Zur frage der herkunft des gasteiner thermal wasser (On the question of the origin of the Bad Gastein thermal waters), English summary. Steirische Beitr. Hydrogeol., v. 21, p. 51-115.

Kusakabe, M., and Wada, H., 1968: Oxygen isotope study of volcanic gases and waters related to Nasudake, Japan. Volcanol. Soc. Bull., v. 13, n. 1, p. 56-57.

Kusakabe, M., Wada, H., and Horibe, Y., 1970: Oxygen and hydrogen isotope ratios of monthly collected waters from Nasudake volcanic area, Japan. J. Geophys. Res., v. 75, n. 30, p. 5941-5951.

Kusakabe, M., 1974: Sulphur isotopic variations in nature; 10, oxygen and sulphur isotope study of Wairakei geothermal well discharges. New Zealand J. Sci., v. 17, n. 2, p. 183-191.

Longinelli, A., 1968: Oxygen isotopic composition of sulfate ions in water from thermal springs. Earth Planet. Sci. Lett., v. 4, n. 3, p. 206-210.

McDonald, D. C., 1966: Hydrogen and oxygen isotopic ratios in the waters of the Ngahwa geothermal region, north Auckland, New Zealand. Bull. Volcanol., v. 29, p. 691-707.

Matsubaya, O., Sakai, H., Kusachi, I., et al., 1973: Hydrogen and oxygen isotopic ratios and major element chemistry of Japanese thermal water systems. Geochem. J., Geochem. Soc. Japan, v. 7, n. 3, p. 123-151.

Matsuo, S., Kusakabe, M., Hirano, M., et al., 1972: D/H and O^{18}/O^{16} ratios of waters from Hakone geothermal system, Japan, abs. Volcanol. Soc. Japan, Bull., v. 17, n. 1, p. 37-38.

Mazor, E., 1973: Paleotemperatures and other hydrological parameters deduced from noble gases dissolved in ground waters, Jordan rift valley, Israel. Geochim. Cosmochim. Acta, v. 36, n. 12, p. 1321-1336.

Mazor, E., Kaufman, A., and Carmi, I., 1973: Hammat Gader, Israel, geochemistry of a mixed thermal spring complex. J. Hydrol., v. 18, n. 3-4, p. 289-303.

Mazor, E., Verhagen, B. T., and Negreanu, E., 1974: Hot springs of the igneous terrain of Swaziland, their noble gases, hydrogen, oxygen and carbon isotopes and dissolved ions (in Isotope Techniques in Ground Hydrology). Vienna, IAEA.

Menendez, R., 1969: Analyse isotopique du deutérium et de l'oxygène dans quatorze eaux de sources thermominérales du bassin Aquitain (Isotopic analysis of deuterium and oxygen in 14 thermomineral springs of the Aquitaine basin). English summary. Cent. Rech. Pau, Bull., v. 3, n. 1, p. 167-185.

Mizutani, Y., and Rafter, T. A., 1969: Oxygen isotope composition of sulphates; pt. 3, Oxygen isotopic fractionation in the bisulphate ion-water system. New Zealand J. Sci., v. 12, n. 1, p. 54-59.

Mizutani, Y., 1972: Estimation of underground temperature from the oxygen isotopic fractionation between sulfate ion and water in geothermal waters, abs. Volcanol. Soc. Japan, Bull., v. 17, n. 1, p. 42.

Wairakei, New Zealand

Mizutani, Y., 1972: Isotopic composition and underground temperature of the Otake geothermal water, Kyushu, Japan. Geochem. J., Geochem. Soc. Japan, v. 6, n. 2, p. 67-73.

Mizutani, Y., and Mamasuna, T., 1972: Origin of the Shimogamo geothermal brine, Izu. Volcanol. Soc. Japan Bull., v. 17, n. 3, p. 123-134.

Murozumi, M., 1961: Isotopic composition of hydrogen and variation of volcanic hydrothermal activity (Japanese with English summary). Volcanol. Soc. Japan, Bull., v. 6, n. 1, Ser. 2, p. 42-46.

Oana, S. : Concentration of heavy water in thermal springs. Balneol. Soc. Japan, J., v. 3, p. 20.

Oana, S., 1942: Heavy water in the thermal waters of Kusatsu. Deutsch. Gesell. Natur. Volkerk. Ostasiens Mitt., v. 33, pt. C, p. 15-19.

Pinneker, E. V., and Lomonosov, I. S., 1973: On genesis of thermal waters in the Sayan-Baikal highland (in Symposium on Hydrogeochemistry and Biogeochemistry, v. 1). Washington, D.C., Clarke Co., p. 246-253.

Reed, M. J., 1975: Water geochemistry, Surprise Valley, California. Calif. Div. Oil and Gas, Geothermal Hot Line, v. 5, n. 1.

Sakai, H., and Matsubaya, O., 1973: Isotope geochemistry of the thermal waters of Japan and its implication to Kuroko ore deposits, abs. Geol. Soc. Am., Abstr., v. 5, n. 7, p. 792.

Sakai, H., and Matsubaya, O., 1974: Isotopic geochemistry of the thermal waters of Japan and its bearing on the Kuroko ore solutions (in Stable Isotopes as Applied to Problems of Ore Deposits). Econ. Geol., v. 69, n. 6, p. 974-991.

Sharma, T., and Pillai, N. V., 1972: Oxygen isotope studies on thermal spring waters from Rajghir and Vajreshwari. Geol. Soc. India, J., v. 13, n. 4, p. 406-409.

Suzuoki, T., Kusakobe, M., and Wada, H., 1967: On the O^{18}/O^{16} and C^{13}/C^{12} ratio in fumarole gases, abs. Volcanol. Soc. Japan, Bull., v. 12, n. 2, p. 97.

Suzuoki, T., 1968: O^{18}/O^{16} ratio of fumarolic water vapor, abs. Volcanol. Soc. Japan, Bull., v. 13, n. 1, p. 57.

Teis, R. V., 1951: Izotopnyi metod apredeleniya temperatur obrazovaniya karbonathykh mineralov. Akad. Nauk SSSR, Doklady, T. 79, n. 2, p. 291-294.

Tongiorgi, E., ed., 1967: Nuclear geology on geothermal areas, Spoleto, 1963. Pisa, Italy, Lab. Geol. Nucleare, 284 p.

Vetshteyn, V. Ye., Lomonosov, I. S., Malyuk, G. A., et al., 1973: Prichiny geograficheskogo raspredeleniya kisloroda-18 i deyteriya v termal'nykh vodakh Sayano-Baykal'skoy gornoy strany (Reasons for the geographical distribution of oxygen-18 and deuterium in the thermal waters of the Sayan-Baikal mountain region). Akad. Nauk SSSR, Izv. Ser. Georg., n. 5, p. 122-128.

White, D. E., and Craig, H., 1959: Isotope geology of the Steamboat Springs area, Nevada, abs. Geol. Soc. Am., Bull., v. 70, n. 12, pt. 2, p. 1696.

White, D. E., 1967: Some principles of geyser activity mainly from Steamboat Springs, Nevada. Am. J. Sci., v. 265, n. 8, p. 641-684.

White, D. E., Barnes, I., and O'Neil, J. R., 1973: Thermal and mineral water of nonmeteoric origin, California coast ranges. Geol. Soc. Am., Bull., v. 84, n. 2, p. 547-559.

13. HYDROCARBONS

Chekalyuk, E. B., and Stefanik, Yu. V., 1972: Rivnovaga stabil'nikh izotopiv vodnyu v sistemi voda-metan (Equilibrium of stable hydrogen isotopes in the water-methane system), Russian summary. Geol. Geokhim. Kopalin (Akad. Nauk Ukr. RSR), n. 31, p. 3-7.

Donovan, T. J., 1974: Petroleum microseepage at Cement, Oklahoma; evidence and mechanism. Am. Assoc. Pet. Geol., Bull., v. 58, n. 3, p. 429-446.

Grinberg, I. V., Petrikovskaya, M. Ye., and Aref'yev, N. V., 1961: K issledovaniyu khimiko-geneticheskikh i izotopnykh sootnosheniy gazovo-kondensatnykh uglevodorodov karpatskogo regiona (Investigation of the chemical-genetic and isotopic ratios of the gas-condensate hydrocarbons of the Carpathian region). L'vov. Geol. Obshch., Geol. Sbornik, n. 7-8, p. 54-65.

Grinberg, I. V., and Petrikovskaya, M. Ye., 1965: Issledovaniye izotopnogo sostava goryuchikh iskopayemykh (Investigation of the isotopic composition of combustible minerals). Kiev, Akad. Nauk Ukrain SSR Inst. Geologii i Geokhimii Goryuchikh Iskopayemykh, 148 p.

Kolodny, Ye., and Gross, S., 1974: Thermal metamorphism by combustion of organic matter; isotopic and petrological evidence. J. Geol., v. 82, n. 4, p. 489-506.

Israel

Kondrat'yeva, G. F., 1968: Improved method of purification of ignition water of petroleum preliminary to measurement of its density by the float method. Geochem. Int., v. 4, n. 4, p. 722-724.

USSR

Libby, L. M., 1972: Multiple thermometry in paleoclimate and historic climate. J. Geophys. Res., v. 77, n. 23, p. 4310-4317.

Maksimov, S. P., Yeremenko, N. A., Pankino, R. G., et al., 1972: Isotopic composition of hydrogen in Soviet oils. Pet. Geol., v. 10, n. 6, p. 231-237.

Mekhtiyev, Sh. R., Bryzgunov, V. S., Vlasova, L. S., Rachinskiy, M. Z., and Soyfer, V. N., 1970: Izotopny sostov vodoroda vod i neftey Apsheronskoy neftegazonosnoy oblasti (Hydrogen isotopic composition of water and oils of the Apsheron oil and gas district). Vyssh. Ucheb. Zavedeniya Izv., Nefti Gaz, n. 1, p. 3-6.

Morales, P., Gaona, S., and Guadalupe, G. M., 1972: Aplicacion de las relaciones isotopicas del oxigeno en carbonatos y de carbono en petroleo (Application of the isotopic relationships of oxygen in carbonates and carbon in petroleum, abs). Union Geofis. Mex. Reun. Anu., Programa Resumenes, n. 1972, p. 39.

Mzhachikh, K. I., and Ashirov, K. B., 1961: K geokhimii deyteriya v neftyakh i bitumakh neftyanogo ryada (On the geochemistry of deuterium in oils and bitumens of the oil group). Sovetskaya Geologiya, n. 6, p. 130-134.

USSR

Nissenbaum, A., 1974: Deuterium content of humic acids from marine and non-marine environments. Mar. Chem., v. 2, n. 1, p. 59-63.

Pankina, R. G., 1967: Izotopnyy sostav sery i vodoroda neftey kak pokazatel' nekotorykh paleogeokhimicheskikh osobennostey neftematerinskogo basseyna (Isotopic composition of the sulfur of some paleogeochemical features of a petroleum parent basin), English summary. Geokhim., n. 5, p. 583-586.

Petrikovskaya, M. Ye., Ivanov, A. K., Kushniruk, V. A., et al., 1969: Issledovaniye izotopnogo sostava metanovykh gazov Mezhrechenskogo kamennougol'nogo mestorozhdeniya v svyazi s yego gazonosnost'yu (Isotope composition of methane from the Mezhrechye coal deposit). Geol. Geokhim. Goryuch. Iskop., Akad. Nauk Ukr. SSR, n. 18, p. 38-45.

Roth, E., 1956: Composition isotopique de l'hydrogène des gaz de Lacq (Isotopic composition of the hydrogen of the gases at Lacq). Acad. Sci. Paris Comptes Rendus, tome 242, n. 26, p. 3097-3100.

Trofimuk, A. A., Cherskiy, N. V., and Tsarev, V. P., 1974: Mekhanizm razdeleniya izotopov vody i gazov v zonakh gidratoobrazovaniya zemnoy kory (Separation mechanism of water and gas isotopes in hydrate-forming zones of the crust). Akad. Nauk SSSR, Doklady, v. 215, n. 5, p. 1226-1229.

Voytov, G. I., Grechukhina, T. G., Lebedev, V. S., et al., 1972: O nekotorykh osobennostyakh khimicheskogo i izotopnogo sostava gazov i vod yuzhnogo Dagestana (Some characteristics of chemical and isotopic composition of gases and waters in south Dagestan). Akad. Nauk SSSR, Doklady, v. 205, n. 5, p. 1217-1220.

Voytov, G. I., et al., 1973: Certain chemical and isotopic characteristics of gases and waters in southern Dagestan, abs. Int. Geol. Rev., v. 15, n. 2, p. 232.

Yapp, C. J., et al., 1975: The determination of the D/H ratio of nonexchangeable hydrogen in cellulose extracted from aquatic and land plants. EOS, v. 56, n. 60, p. 480.

Yeremenko, N. A., Botneva, T. A., Maksimov, S. P., et al., 1971: Variatsii stabil'nykh izotopov ugleroda, vodoroda i sery neftey v svyazi s tsiklichnost'yu protsessov neftegazoobrazovaniya (Variations of the carbon, hydrogen, and sulfur isotopes in connection with the processes of oil and gas formation). Geol. Nefti. Gaza, n. 4, p. 30-33.

14. PALEOTEMPERATURES AND PALEOCLIMATOLOGY

A. GENERAL

Bray, J. R., 1971: Solar-climate relationship in the post-Pleistocene. Science, v. 171, n. 3977, p. 1242-1243.

Broecker, W. S., and van Donk, J., 1969: Insolation changes, ice volumes and the O^{18} record in deep-sea cores. Lamont-Doherty Geological Observatory (NTIS), 34 p.

DeVries, H., 1957: Contribution of radiocarbon dating and measurement of paleotemperatures to Pleistocene correlations. Geologie en Mijnbouw (NW Ser), 19e Jaargang, p. 303-304.

Emiliani, C., 1967: Isotope paleotemperatures, reply to discussion of paper by C. Emiliani, 1966). Science, v. 157, n. 3789, p. 723-725.

Faegri, K., 1970: Temperaturkurver for de siste 10,000 ar (Temperature curves for the last 10,000 years). Naturen, v. 94, n. 7, p. 431-435.

Fontes, J. C., Létolle, R., Faure, H., et al., 1971: Paléoclimatologie isotopique et paléoécologie du littoral holocene de Mauritanie (Isotopic paleoclimatology and paleoecology of the Holocene coast of Mauritania, abs). (in Les Niveaux Marins Quarternaires, pt. 1, Holocene). Quaternaria, v. 14, p. 209-210.

Frakes, L. A., and Kemp, E. M., 1972: Influence of continental positions on early Tertiary climates. Nature, v. 240, n. 5376, p. 97-100.

Frakes, L. A., and Kemp, E. M., 1973: Paleocene continental positions and evolution of climate; with comment and reply (in Implications of Continental Drift to the Earth Sciences). London, Acad. Press, v. 1, p. 539-559.

Gill, E. D., 1972: Paleoclimatology and dinosaurs in southeast Australia. Search, v. 3, n. 11-12, p. 444-446.

Gorbarenko, S. A., 1971: Opredeleniya klimaticheskikh usloviya Pleystotsena metodom izotopnoy paleotermometrii (Determining Pleistocene climatic conditions by the method of isotope paleothermometry). Moscow Univ., Vestn., Ser. Geogr., v. 26, n. 5, p. 106-109.

Létolle, R., 1972: Les "paléotempératures" isotopiques problèmes et perspectives (Isotopic paleotemperatures, problems and perspectives) (in Colloque sur les Méthodes et Tendances de la Stratigraphie), English summary. Fr. Bur. Rech. Géol. Minières, Mem., n. 77, v. 2, p. 753-758.

Lowenstam, H. A., 1968: Paleotemperatures of the Permian and Cretaceous periods (in Problem in Paleoclimatology-NATO Palaeoclimates Conf., Univ. New Castle-upon-Tyne 1963 Proc.). London-New York, Interscience Publishers, p. 227-248, 251-252.

Naidin, D. P., 1972: Isotopic paleotemperatures and some problems of geology. Byul. Moskov. Obschest. Ispytatelei Prirody, Otd. Geol., v. 47, n. 5, p. 112-124.

Onuma, N., Clayton, R. N., and Mayeda, T. S., 1974: Oxygen isotope cosmothermometer revisited (letter). Geochim. Cosmochim. Acta, v. 38, n. 1, p. 189-191.

Shackleton, N. J., 1972: Ocean paleotemperature changes around Africa. Palaeoecol. Afr., v. 6, p. 37-38.

Shackleton, N. J., 1973: Oxygen isotope palaeoclimatology, abs. EOS, Am. Geophys. Union, Trans., v. 54, n. 4, p. 337-338.

Stehli, F. G., 1972: An approach to Permian climate (in International Symposium on the Carboniferous and Permian Systems in South America). San Paulo, Brazil, Acad. Bras. Cienc., San Paulo, Univ., Inst. Geoscienc., p. 44-45.

Tudge, A. P., 1960: A method of analysis of oxygen isotopes in orthophosphate, its use in the measurement of paleotemperatures. Geochim. Cosmochim. Acta, v. 18, n. 1/2, p. 81-93.

Urey, H. C., 1953: The measurement of paleotemperatures, summary (in Proceedings of the Conference on Nuclear Processes in Geologic Settings, September 1953). Natl. Res. Coun. Comm. Nuclear Sci., chap. 20, p. 71-72.

Vinogradov, A. P., 1966: Izotopnyye ravnovesiya i geologicheskoye problemy (Isotopic equilibrium and geologic problems). Akad. Nauk SSSR Izv. Ser. Geol., n. 1, p. 7-16.

Westgate, J. A., Fritz, P., Matthews, J. V., Jr., et al., 1972: Geochronology and palaeoecology of mid-Wisconsin sediments in west-central Alberta, Canada, abs. Int. Geol. Congr. Abstr., Congr. Geol. Int., Resumes, n. 24, p. 380.

B. PEAT, FOSSIL WOOD AND POLLEN

Epstein, S., and Yapp, C. J., 1975: Climatic implications of the D/H ratio of hydrogen in C-H groups in tree cellulose. EOS, v. 56, n. 6, p. 480.

Arizona, California, Scotland

Schiegl, W. E., 1972: Deuterium content of peat as a paleoclimatic recorder. Science, v. 175, n. 4021, p. 512-513.

Netherlands

Schiegl, W. E., and Trimborn, P., 1973: Die palaeoklimatische bedeutung des deuteriumgehaltes von fossilen holzproben aus den baendertonen von Baumkirchen, Inntal (The paleoclimatological significance of the deuterium content of fossil wood samples in the laminated silt and clay of Baumkirchen, Inn Valley, Austria). A. Gletscherk. Glazialgeol., v. 8, n. 1-2, p. 231-233.

Schiegl, W. E., 1974: Deuterium-thermometer in fossilen pflanzen (The deuterium thermometer in fossil plants), English summary. Umschau, v. 74, n. 13, p. 421-422.

Schiegl, W. E., 1974: Climatic significance of deuterium abundance in growth rings of Picea. Nature, v. 251, n. 5476, p. 582-584.

Germany

C. CARBONATES

General

Bagge, E., 1954: Isotopen-bestimmungen als hilfsmittel paleontologischer forschung (Isotope determinations as an aid to paleontological research). Umschau, Jahrg. 54, Heft 12, p. 364-366.

Bowen, R., 1966: Paleotemperature analysis. New York, Elsevier Publishing Co., 265 p.

Bowen, R., 1966: Oxygen isotopes as climatic indicators. Earth Sci. Rev., v. 2, n. 3, p. 199-224.

Dorman, F. H., 1968: Some Australian oxygen isotope temperatures and a theory for a 30-million-year world temperature cycle. J. Geol., v. 76, n. 3, p. 297-313.

Antarctica

Emiliani, C., 1966: Isotopic paleotemperatures. Science, v. 154, n. 3751, p. 851-857.

Fabricius, F., Friedrichsen, H., and Jacobshagen, V., 1970: Zur methodik der palaeotemperaturermittlung in Obertrias und Lias der Alpen und benachbarter Mediterrean-gebiete (Methods for the determination of paleotemperatures in the upper Triassic and Liassic of the Alps and the neighboring region) (in Rezente und Fossile Karbonat-Sedimentation ein Symposium, Austria), English summary. Geol. Bundesanst. Verh., n. 4, p. 583-593.

Folinsbee, R. E., Fritz, P., Krouse, H. R., et al., 1970: Carbon-13 and oxygen-18 in dinosaur, crocodile, and bird eggshells indicate environmental conditions. Science, v. 168, n. 3937, p. 1353-1356.

Gorsline, D. S., and Barnes, P. W., 1972: Carbonate variations as climatic indicators in contemporary California Flysch basins (in Stratigraphy and Sedimentology, Sec. 6). Internat. Congr., Proc., Congr. Geol. Int., Programme, n. 24, p. 270-277.

Létolle, R., DeLumley, H., and Grazzini, C., 1971: Isotopic composition of quaternary organic carbonates of the western Mediterranean; attempt of climatic interpretation. Comptes Rendus, Sér. D, v. 273, n. 23, p. 2225-2228.

Lowenstam, H. A., and Epstein, S., 1954: Paleotemperatures of the post-Aptian Cretaceous as determined by the oxygen isotope method. J. Geol., v. 62, n. 3, p. 207-248.

Schoell, M., Morteani, G., and Hoermann, P. K., 1973: Relationship between carbonate mineralization and regional metamorphism in the western Tavern area as deduced from C^{13} and O^{18} investigations on the carbonates, abs. (in Europa isches Kolloquium fuer Geochronologie, II). Referate, Fortschr. Mineral., v. 50, Beih. 3, p. 126.

Stuiver, M., 1968: Oxygen-18 content of atmospheric precipitation during last 11,000 years in the Great Lakes region. Science, v. 162, n. 3857, p. 994-997.

Stuiver, M., 1970: Oxygen isotope ratio of fresh water carbonates as a climatic indicator, abs. (in American Quaternary Association, 1st Meeting). Seattle, Washington, Am. Quat. Assoc., p. 128.

Stuiver, M., 1970: Oxygen isotope ratios of fresh water carbonates as climatic indicators. J. Geol. Res., v. 75, n. 27, p. 5247-5257.

Maine, New York, Indiana, South Dakota, Connecticut

Teis, R. V., Chupakhin, M. S., and Naidin, D. P., 1960: Determination of paleotemperatures according to the composition of oxygen of organogenous calcite. English summary. Doklady Sov. Geol. Problema 1, p. 146-156.

Crimea

Teis, R. V., Naidin, D. P., and Zadorozhnyy, I. K., 1969: Paleotemperatures of the upper Cretaceous of the Russian platform and other parts of the Soviet Union from the isotopic composition of oxygen in organogenic calcite.

(in Problems of Geochemistry, N. I. Khitarov, ed.). Jerusalem, Israel Program Sci. Transl., p. 704-720.

Weber, J. W., 1964: Variations of oxygen-18/oxygen-16 and carbon-13/carbon-12 in the calcium carbonate Pleistocene varved clays from Toronto, Canada. Nature, v. 202, n. 4934, p. 791-792.

Weber, J. N., 1964: Oxygen isotope ratios in fresh water limestones as sensitive paleoclimate temperature indicators, abs. Am. Assoc. Petrol. Geol. Bull., v. 48, n. 4, p. 551.

Weber, J. N., 1969: Carbon and oxygen isotope composition of carbonate material in the Rita Blanca varves (in Paleoecology of an Early Pleistocene Lake on the High Plains of Texas). Geol. Soc. Am. Mem. 113, p. 47-58.

Travertine

Duplessy, J. C., Lebeyrie, J., Lalou, C., and Nguyen, H. V., 1970: Continental climatic variations between 130,000 and 90,000 years BP. Nature, v. 226, n. 5246, p. 631-633.

Duplessy, J. C., Lalou, C., and Gomes de Azeu edo, A. E., 1969: Etude des conditions de concrétionnement dans les grottes au moyens des isotopes stables de l'oxygène et du carbone (Conditions of concretions in caves studied by means of stable isotopes of carbon and oxygen). Acad. Sci., Comptes Rendus, Sér. D, v. 268, n. 19, p. 2327-2330.

Duplessy, J. C., Lalou, C., Delibrias, G., et al., 1972: Datations et études isotopiques des stalagmites; applications aux paléotempératures (Dating and isotopic studies of stalagmites; interpretation of paleotemperatures), English summary. Ann. Speleol., v. 27, n. 3, p. 445-461.

Duplessy, J. C., Labeyre, J., Lalou, C., and Nguyen, H. V., 1971: La mésure des variations climatiques continentales; application à la période comprise entre 130,000 et 90,000 ans BP (The measurement of continental climatic variations; application to the period between 130,000 and 90,000 years BP), English abstract. Quaternary Res., v. 1, n. 2, p. 162-174.

Fornaca-Rinaldi, G., Panichi, C., and Tongiorgi, E., 1968: Some causes of the variation in the isotope composition of carbon and oxygen in cave concretions. Earth Planet. Sci. Lett., v. 4, n. 4, p. 321-324.

Fritz, P., 1965: Composizione isotopica dell'ossigeno e del carbonio nei travertini della Toscona (Oxygen and carbon isotope composition in the travertines from Tuscany), English abstract. Boll. Geofisica Teor. ed Appl., v. 7, n. 25, p. 25-30.

Geyh, M. A., 1970: Isotopenphysikalische untersuchungen an kalksinter, ihre bedeutung fuer die C^{14} altersbestimmung von grundwasser und die erforschung des palaeoklimas (Isotope-physical studies of travertine, their importance to the C-14 age determination of ground water and to paleoclimatological research), English, French and Russian summaries. Ger., Bundesanst. Bodenforsch., Geol. Jahrb., v. 88, p. 149-158.

Thompson, P., 1971: Oxygen isotopes of speleothems as an aid to paleoclimatic determination in the Appalachians, abs. Caves Karst, v. 13, n. 6, p. 50.

Thompson, P., Schwarcz, H. P., and Ford, D. C., 1974: Continental Pleistocene climatic variations from Speleothem Age and isotopic data. Science, v. 184, n. 4139, p. 893-895.

Thompson, P., Schwarcz, H. P., and Ford, D. C., 1972: Isotopic dating and paleoclimate studies of cave deposits in West Virginia, abs. Geol. Soc. Am., Abstr., v. 4, n. 7, p. 689.

Marine Sediments

Coplen, T. B., and Schlanger, S. O., 1973: Oxygen and carbon isotope studies of carbonate sediments from site 167, Magellan Rise, leg 17 (in Initial Reports of the Deep Sea Drilling Project). Washington, D.C., U.S. Govt. Printing Office, v. 17, p. 505-509.

Donn, W., Shaw, D. M., and Emiliani, C., 1967: Isotopic paleotemperatures--discussion of paper by C. Emiliani, 1966, and reply. Science, v. 157, n. 3789, p. 722-725.

Emiliani, C., 1957: Oxygen isotope measurements of deep sea sediments. Washington, D.C., Natl. Res. Coun. Pub. 473, p. 67-78.

Atlantic Ocean

Emiliani, C., and Mayeda, T., 1961: Carbonate and oxygen isotopic analysis of core 241A. J. Geol., v. 69, n. 6, p. 729-732.

Emiliani, C., 1963: The significance of deep sea cores (in Science in Archaeology--A Comprehensive Survey of Progress and Research). New York, Basic Books, p. 99-107.

Emiliani, C., 1964: Paleotemperature analysis of the Caribbean cores A254-BR-C and CP-28. Geol. Soc. Am. Bull., v. 75, n. 2, p. 129-144.

Emiliani, C., 1966: Oxygen isotopes, oceanic sediments, and the ice ages, abs. Am. Geophys. Union, Trans., v. 47, n. 1, p. 44.

Emiliani, C., 1966: Paleotemperature analysis of Caribbean core P6304-8 and P6304-9 and a generalized temperature curve for the past 425,000 years. J. Geol., v. 74, n. 2, p. 109-126.

Emiliani, C., 1971: The last interglacial-paleotemperatures and chronology. Science, v. 171, n. 3971, p. 571-573.

Emiliani, C., 1972: Quaternary paleotemperatures and the duration of the high-temperature intervals. Science, v. 178, n. 4059, p. 398-401.

Emiliani, C., and Shackleton, N. J., 1974: The Brunhes epoch; isotopic paleotemperatures and geochemistry. Science, v. 183, n. 4124, p. 511-514.

Atlantic Ocean; Caribbean Sea

Herman, Y., 1972: South Pacific quaternary paleo-oceanography, abs. Int. Geol. Congr. Abstr., Congr. Geol. Int., Resumes, n. 24, p. 260.

Kemp, W. C., and Eger, D. T., 1967: The relationships among sequences with applications to geological data. J. Geophys. Res., v.77 2, n. 2, p. 739-751.

Caribbean Sea

Lloyd, R. M., 1973: Preliminary isotopic investigations of samples from deep sea drilling in the Mediterranean Sea, DSDP leg 13 (in Initial Reports of the Deep Sea Drilling Project). Washington, D.C., U.S. Govt. Printing Office, v. 13, pt. 2, p. 783-787.

Caribbean Sea

Rosholt, J. N., Jr., et al., 1962: Pa^{231}/Th^{230} dating and O^{18}/O^{16} temperature analysis of core A254-BR-C. J. Geophys. Res., v. 67, n. 7, p. 2907-2911.

Savin, S. M., Douglas, R. G., and Stehli, F. G., 1972: Oxygen studies, paleotemperature studies of Tertiary ocean sediments, abs. Geol. Soc. Am., Abstr., v. 4, n.7, p. 653-654.

Fossils

General

Anglada, R., Granier, J., Létolle, R., et al., 1972: Variations climatiques dans le Pliocene inférieur des Restanques, Nice, Alpes-Maritimes (Climatic variations in the lower Pliocene of Restanques, Nice, Alps-Maritime). Acad. Sci. Comptes Rendus, Ser. D, v. 275, n. 25, p. 2845-2848.

Allen, P., Keith, M. L., Tan, F. C., et al. : Isotopic ratios and Wealden environments. Palaeontology, v. 16, pt. 3, p. 607-621.

England

Cortecci, G., and Longinelli, A., 1973: O^{18}/O^{16} ratios in sulfate from fossil shells. Earth Planet. Sci. Lett., v. 19, n. 4, p. 410-412.

Duplessy, J. C., and Labeyrie, L., 1973: Comparison de la composition isotopiques des forminifères et des diatomées en milieu océanique (Comparison of the isotopic composition of foraminifera and diatoms in a marine environment, abs.) Réun. Annu. Sci. Terre, Programme Résumés, p. 174.

Durazzi, J. T., 1973: Magnesium, strontium and oxygen isotopes in the shells of ostracods, abs. Geol. Soc. Am., Abstr., v. 5, n. 7, p. 606.

Emiliani, C., 1958: Ancient temperatures. Sci. Am., v. 198, n. 2, p. 54-63.

Forester, R. M., Sandberg, P. A., and Anderson, T. F., 1973: Isotopic variability of cheilostome Bryozoan skeleton (in Living and Fossil Bryozoa, Internat. Bryozoology Assoc. Conf., 2nd Proceedings). London, Acad. Press, p. 79-94.

Hathaway, J. C., and Degens, E. T., 1969: Methane-derived marine carbonates of Pleistocene age. Science, v. 165, n. 3894, p. 690-692.

Atlantic Ocean

Hoefs, J., and Sarnthein, M., 1971: O^{18}/O^{16} ratios and related temperatures of recent pteropod shells (*Cavolinia longirostris leseur*) from the Persian Gulf. Marine Geol., v. 10, n. 4, p. M11-M16, 6 p.

Mook, W. G., and Vogel, J. C., 1968: Isotopic equilibrium between shells and their environment. Science, v. 159, n. 3817, p. 874-875.

Mook, W. G., 1971: Paleotemperatures and chlorinites from stable carbon and oxygen isotopes in shell carbonates. Palaeogeogr. Palaeoclimatol. Palaeoecol., v. 9, n. 4, p. 245-263.

Netherlands

Shackleton, N., and Renfrew, C., 1970: Neolithic trade routes realigned by oxygen isotope analyses. Nature, v. 228, n. 5276, p. 1062-1065.

Sisler, F. D., 1959: Biogeochemical concentration of deuterium in the marine environment, abs. Science, v. 129, n. 3358, p. 1288.

Sisler, F. D., 1960: Geomicrobiological effect on hydrogen-isotope equilibria in the marine environment, abs. Geol. Soc. Am. Bull., v. 71, n. 12, pt. 2, p. 1974.

Tan, F. C., and Hudson, J. D., 1971: Isotopic composition of carbonates in a marginal marine formation. Nature, Phys. Sci., v. 232, n. 30, p. 87-88.

Scotland

Tan, F. C., and Hudson, J. D., 1974: Isotopic studies of the palaeoecology and diagenesis of the Great Estuarine Series (Jurassic) of Scotland. Scot. J. Geol., v. 10, pt. 2, p. 91-128.

Togliatti, V., 1968: Il rapporto O-18/O-16 in *Patella coerulea* della costa Tirrenica (The O-18/O-16 ratio in *Patella coerulea* on the Tyrrhenian coast). Accad. Sci. lst Bologna Atti, Cl. Sci. Fiz. Rend. 1965-66, Ser. 12, v. 3, n. 1-2, p. 70-82.

Italy

Weber, J. N., 1964: Carbon and oxygen isotope ratios as environmental indicators; anomalous results from carbonate shells from beach sediments of Lake Managua, Nicaragua. Nature, v. 201, n. 4914, p. 63.

Coral Reefs

Devereux, I., 1967: Oxygen isotope palaeotemperature measurements on two Tertiary deep-water coral thickets from Wairarapa, New Zealand. Palaeogeogr. Palaeoclimatol. Palaeoecol., v. 3, n. 4, p. 447-455.

Keith, M. L., and Weber, J. N., 1965: Systematic relationships between carbon and oxygen isotopes in carbonates deposited by modern corals and algae. Science, v. 150, n. 3695, p. 498-501.

Jamaica

Ma, T.-Y. H., 1964: A comparison of the study of Upper Jurassic climate based on growth values of reef corals with that by the oxygen isotope method (in Studies on Oceanography--A Collection of Papers Dedicated to Koji Hidaka). Tokyo, University Press, p. 496-514.

Weber, J. N., and Woodhead, P. M. J., 1972: Temperature dependence of oxygen-18 concentration in reef coral carbonates. J. Geophys. Res., v. 77, n. 3, p. 463-473.

Weber, J. N., 1973: Deep sea ahermatypic scleractinian corals; isotopic composition of the skeleton. Deep Sea Res., v. 20, n. 10, p. 901-909.

Woodhead, R. M. J., and Weber, J., 1972: Evolution of the reef-building corals and the significance of their association with zooanthellae, abs. Int. Symp. Hydrogeochem. Biogeochem., Abstr., p. 85.

Woodhead, P. M. J., and Weber, J. N., 1973: The evolution of reef-building corals and the significance of their association with zooxanthellae (in Symposium on Hydrogeochemistry and Biogeochemistry, Proceedings, v. 2). Washington, D.C., Cla ke Co., p. 280-304.

Australia

Foraminifera

Allegre, C., and Javoy, M., 1964: Utilization des isotopes de l'oxygène en paléogéographie (The use of oxygen isotopes in paleogeography) (in Paleontology and Stratigraphy, 22nd Int. Geol. Congr., India). Int. Geol. Congr. Rep., n. 22, pt. 8, p. 438-444.

Allegre, C., Boulanger, D., and Javoy, M., 1963: Étude à l'aide des isotopes l'oxygène de la paléothermométrie du Nummulitique basque (Study of the paleothermometry of the Nummilitic deposits of the Basque region by means of oxygen isotopes). Soc. Geol. France, Comptes Rendus, n. 8, p. 256-257.

Ault, W. U., 1959: Oxygen isotope measurements on Arctic cores. U.S. Air Force Cambridge Research Center, Geophys. Res. Papers, n. 63, Scientific Studies at Fletcher's Ice Island, T-3 (1952-55), v. 1, p. 159-168.

Bandy, O. L., 1972: A review of the calibration of deep sea cores based upon species variation, productivity, and O^{18}/O^{16} ratio of planktonic foraminifera, including sedimentation rates and climatic inferences; with discussion (in Calibration of Hominoid Evolution). Edinburgh, Scot. Acad. Press, p. 37-61.

Barash, M. S., and Gromova, T. S., 1969: Paleotemperaturnyy analiz kolonok po planktonnym foraminiferam (Paleotemperature analysis of cores, based on planktonic foraminifera) (in Osnovnyye Problemy Mikropaleontologii i Organogennogo Osadkonakopleniya v Okeanakh i Moryakh; K VIII Kongressu INQUA), English summary. Akad. Nauk SSSR, Okeanogr. Komm., Moscow, p. 153-169.

Barash, M. S., Nikolayev, S. D., and Blyum, N. S., 1973: Paleotemperaturnyy analiz trekh kolonok osadkow severnoy Atlantiki (Paleotemperature analysis of three sediment cores), English summary. Akad. Nauk SSSR, v. 13, n. 6, p. 1052-1058.

Bé, A. W. H., and von Donk, J., 1971: Oxygen-18 isotopes of recent planktonic foraminifera. Science, v. 173, n. 3993, p. 167-168.

Bellaiche, G., Gaundry, I., and Vergnaud, G. C., 1971: Paléogéographie quatérnaire du golfe de Frejus, Var (Quaternary paleogeography of the Gulf of Frejus, Var). (in Etudes sur le Quaternaire dans le Monde, v. 1). Assoc. Fr. Étude Quat., Bull., Suppl. n. 4, p. 165-178.

Mediterranean

Douglas, R. G., and Savin, S. M., 1971: Isotopic analysis of planktonic foraminifera from the Cenozoic of the northwest Pacific, leg 6 (in Initial Reports of the Deep Sea Drilling Project, v. 6, leg 6 of Glomar Challenger, Honolulu, Hawaii to Apra, Guam, June-August 1969). Washington, D.C., U.S. Govt. Printing Office, p. 1123-1127.

Douglas, R. G., and Savin, S. M., 1973: Oxygen and carbon isotope analysis of Cretaceous and Tertiary foraminifera from the central north Pacific, DSDP leg 17 (in Initial Reports of the Deep Sea Drilling Project). Washington, D.C., U.S. Govt. Printing Office, v. 17, p. 591-603.

Duplessy, J. C., Lalou, C., and Vinot, A. C., 1969: Differential isotopic fractionation in benthic foraminifera and paleotemperatures reassessed. Science, v. 168, n. 3928, p. 250-251.

Atlantic Ocean

Duplessy, J. C., and Vinot, A. C., L'interpretation des variations de la composition isotopique des tests de foraminifères durant le Pléistocene (Interpretation of isotopic composition variations in foraminifera tests during the Pleistocene) (in Colloque sur les Méthodes et Tendances de la Stratigraphie), English summary. Fr. Bur. Rech. Géol. Minières, Mém., n. 77, v. 2, p. 759-765.

Emiliani, C., 1955: Pleistocene temperature variations in the Mediterranean. Quaternaria, v. 2, p. 87-98.

Emiliani, C., 1961: The temperature decrease of surface sea water in high latitudes and of abyssal-hadal water in open oceanic basins during the past 75 million years. Deep Sea Res., v. 8, n. 2, p. 144-147.

Emiliani, C., 1971: Depth habitats of growth stages of pelagic foraminifera. Science, v. 173, n. 4002, p. 1122-1124.

Emiliani, C., 1971: Paleotemperature variations across the Plio-Pleistocene boundary. Science, v. 171, n. 3966, p. 60-62.

Italy

Emiliani, C., 1971: The amplitude of Pleistocene climatic cycles at low latitudes and the isotopic composition of glacial ice (in The Late Cenozoic Glacial Ages, Karl K. Turekian, ed). New Haven, Conn., Yale Univ. Press, p. 183-197.

Emiliani, C., 1974: Isotopic temperatures and shell morphology of *Globigerimoides ruba* in the Mediterranean deep sea core 189. Micropaleontology, v. 20, n. 1, p. 106-109.

Hays, J. D., 1974: Chronology of Ice Age climates; the last million years, abs. EOS, Am. Geophys. Union, Trans., v. 55, n. 4, p. 258.

Hecht, A. D., and Savin, S. M., 1970: Oxygen-18 studies of recent planktonic foraminifera; comparison of phenotypes and of test parts. Science, v. 170, n. 3953, p. 69-71.

Hecht, A. D., and Savin, S. M., 1971: Oxygen-18 isotopes of recent planktonic foraminifera (reply to discussion by A. W. H. Bé and J. von Donk of 1970 paper). Science, v. 173, n. 3992, p. 168-169.

Hecht, A. D., 1973: Faunal and oxygen isotopic paleotemperatures and the amplitude of glacial/interglacial temperature changes in the equatorial Atlantic, Caribbean Sea and Gulf of Mexico. Quat. Res. (Washington Univ., Quat. Res. Cent.), v. 3, n. 4, p. 671-690.

Herman, Y., O'Neil, J. R., and Drake, C. L., 1972: Micropaleontology and paleotemperatures of postglacial SW Greenland fjord cores (in Climatic Changes in Arctic Areas during the Last Ten-Thousand Years). Acta Univ. Ouluensis, Ser. A, Sci. Rerum Nat., n. 3, Geol. n. 1, p. 357-381.

Imbrie, J., 1972: Correlation of the climatic record of the Camp Century Ice Core (Greenland) with foraminiferal paleotemperature curves from North Atlantic deep sea core, abs. Geol. Soc. Am., Abstr., v. 4, n. 7, p. 550.

Imbrie, J., van Donk, J., and Kipp, N. G., 1973: Paleoclimatic investigation of a late Pleistocene Caribbean deep sea core; comparison of isotopic and faunal methods. (in CLIMAP Program). Quat. Res. (Washington Univ., Quat. Res. Cent.), v. 3, n. 1, p. 10-38.

Lidz, B., Kehm, A., and Miller, H., 1968: Depth habitats of pelagic foraminifera during the Pleistocene. Nature, v. 217, n. 5125, p. 245-247.

Caribbean

Oba, T., 1972: O^{18}/O^{16} paleotemperatures (in Prof. Jun-ichi Iwai Memorial Volume), Japanese and English summaries. Sendai, Japan, Tohoku Univ., Inst. Geol. Paleontol., p. 139-145.

Podgoretskiy, V. V., and Popovich, Ye. F., 1968: Izotopnaya paleotermometriya eotsenovogo basseyna nizhnego povolzh'ya i yuzhnykh yergeney po rakovinam krupnykh foraminifer (Isotope paleothermometry of the Eocene basin of the lower Volga and the southern Ergenei according to shells of large foraminifera), English summary. Geokhim. n. 11, p. 1382-1387.

Rotschy, F., Vergnaud Grazzini, C., Bellaiche, G., et al., 1972: Etude paléoclimatologique d'une carotte prélevée sur un dome de la plaine abyssale ligure, "structure Alinat" (Paleoclimatological study of a core from an abyssal hill in the Ligurian Sea, the Alinat structure). Palaeogeogr. Palaeoclimatol. Palaeoecol., v. 11, n. 2, p. 125-145.

Mediterranean Sea

Sackett, W. M., and Rankin, J. D., 1970: Paleotemperatures for the Gulf of Mexico. J. Geophys. Res., v. 75, n. 24, p. 4557-4560.

Saito, T., and van Donk, J., 1974: Oxygen and carbon isotope measurements of late Cretaceous and early Tertiary foraminifera. Micropaleontology v. 20, n. 2, p. 152-176.

Savin, S. M., and Douglas, R. G., 1973: Stable isotope and magnesium geochemistry of recent planktonic foraminifera from the South Pacific. Geol. Soc. Am. Bull., v. 84, n. 7, p. 2327-2342.

Savin, S. M., and Stehl, F. G., 1973: The oxygen isotope composition of the oceans during the past 170,000 years, abs. EOS, Am. Geophys. Union, Trans., v. 54, n. 4, p. 328.

Atlantic Ocean

Shackleton, N., 1967: Oxygen isotope analyses and Pleistocene temperatures re-assessed. Nature, v. 215, n. 5096, p. 15-17.

Caribbean Sea

Smith, P. B., and Emiliani, C., 1968: Oxygen isotope analysis of recent tropical Pacific benthonic foraminifera. Science, v. 160, n. 3834, p. 1335-1336.

van Donk, J., and Mathieu, G., 1969: Oxygen isotope compositions of foraminifera and water samples from the Arctic Ocean. J. Geophys. Res., v. 74, n. 13, p. 3396-3407.

Vergnaud-Grazzini, C., and Bartolini, C., 1970: Evolution paléoclimatique des sédiments wurmiens et post-wurmiens en mer d'Alboran (Paleoclimatic evolution of Würmian and post-Würmian sediments from the sea near Alboran), English summary. Rev. Géogr. Phys. Géol. Dyn., v. 12, n. 4, p. 325-333.

Mediterranean Sea

Mollusks

Emiliani, C., 1956: Oxygen isotopes and paleotemperature determinations. Internat. Quaternary Cong. 4th, Rome-Pisa, 1953, Actes, v. 2, p. 831-844.

Emiliani, C., Mayeda, T., and Selli, R., 1961: Paleotemperature analysis of the Plio-Pleistocene section at Le Castella, Calabria, southern Italy. Geol. Soc. Am. Bull., v. 72, n. 5, p. 679-688.

Emiliani, C., and Mayeda, T., 1964: Oxygen isotope analysis of some molluscan shells from fossil littoral deposits of Pleistocene age. Am. J. Sci., v. 262, n. 1, p. 107-113.

Portugal, Italy, France and Morocco

Gorbarenko, S. A., Nikolayev, S. D., and Popov, S. V., 1973: Izotopnyy sostav kisloroda rakovin chetvertichnykh mollyuskov i izmeneniya paleogeografii Vostochnogo Kaspiya (Isotopic content of oxygen in shells of quaternary mollusks and the evolution of paleogeography of the eastern Caspian region). Mosk. Ovo. Ispyt. Prir. Byull., Otd. Geol., v. 48, n. 3, p. 102-109.

Kaltenegger, W., Preisinger, A., and Roegl, F., 1971: Palaeotemperaturbestimmungen an aragonitschaligen mollusken aus dem alpinen Mesozoikum (Paleotemperature determinations for aragonitic mollusks from the Alpine Mesozoic), English summary. Palaeogeogr. Palaeoclimatol. Palaeoecol., v. 10, n. 4, p. 273-285.

Keith, M. L., and Anderson, G. M., 1963: Isotopic within-shell variation in mollusks in relation to their environment, abs. Geol. Soc. Am. Spec. Paper 73, p. 185.

Keith, M. L., and Parker, R. H., 1965: Local variation of C^{13} and O^{18} content of mollusk shells and the relatively minor temperature effect in marginal marine environments (in Organic Matter in Marine Sediment). Marine Geol., v. 3, n. 1/2, p. 115-129.

Longinelli, A., Cortecci, G., and Fornaca-Rinaldi, G., 1972: Rinvenimento di una linea di spiaggia sepolta di eta' wurmiana al largo del litorale toscano (Evidence of Wurm beach sediments off the Tuscany littoral area), English summary. Soc. Geol. Ital., Boll., v. 91, n. 1, p. 3-10.

Shackleton, N., 1970: Stable isotope study of the palaeoenvironment of the Neolithic site of Nea Nikomedeia, Greece. Nature, v. 227, n. 5261, p. 943-944.

Valentine, J. W., and Meade, R. F., 1961: Californian Pleistocene paleotemperatures. California Univ., Pub. Geol. Sci., v. 40, n. 1, p. 1-45.

Echinoderms

Weber, J. N., 1956: Fractionation of the stable isotopes of carbon and oxygen in marine calcareous organisms--the Echinicidea, part II. Environmental and genetic factors. Geochim. Cosmochim. Acta, v. 30, n. 7, p. 705-736.

Pelecypods

Dodd, J. R., 1966: Diagenetic stability of temperature-sensitive skeletal properties in Mytilus from the Pleistocene of California. Geol. Soc. Am. Bull., v. 77, n. 11, p. 1213-1224.

Dorman, F. H., and Gill, E. D., 1959: Oxygen isotope paleotemperature determinations of Australian Cainozoic fossils. Science, v. 130, n. 3388, p. 1576.

Dorman, F. H., 1966: Australian tertiary paleotemperatures. J. Geol., v. 74, n. 1, p. 49-61.

Keith, M. L., Anderson, G. M., and Eichler, R., 1964: Carbon and oxygen isotope composition of mollusk shells from marine and freshwater environments. Geochim. Cosmochim. Acta, v. 28, n. 11, p. 1757-1786.

Létolle, R., and Tivollier, J., 1966: Rélations entre la paléoécologie et la composition isotopique de l'oxygène et du carbone des lamellibranches du tertiaire parisien (Relationship between paleoecology and the isotopic composition of oxygen and carbon in pelecypod shells from the Tertiary of Paris region). Acad. Sci. Comptes Rendus, Sér. D, v. 263, n. 23, p. 1824-1826.

Lloyd, R. M., 1969: A paleoecological interpretation of the Caloosa Hatcher formation using stable isotopes. J. Geol., v. 77, n. 1, p. 1-25.

Zhirmunskiy, A. V., Zadorozhnyy, I. K., Naydin, P. D., et al., 1968: Determination of temperatures of growth of modern and fossil mollusks by O^{18}/O^{16} ratio of their shells. Geochem. Int., v. 4, n. 3, p. 459-468.

Belemnites

Berlin, T. S., Naydin, D. P., Saks, V. N., Teys, R. V., and Khabakov, A. V., 1966: Klimaty v yurskom i melovym periodakh na severe SSSR po paleotemperaturnym opredeleniyam (Climates in the Jurassic and Cretaceous periods in the northern part of the USSR according to paleotemperature determinations), English summary. Akad. Nauk SSSR Sibirskoye Otdeleniye, Geologiya i Geofizika, n.10, p. 17-31.

Berlin, T. S., and Khabakov, A. V., 1967: Sravneniye rezul'tatov opredeleniy paleotemperatur v moryakh verkhnego mela po indeksam kal'tsiy-magniyevogo otnosheniya i po dannym mass-spektrometrii izotopov kisloroda v rostakh belemnitellid (Comparison of the results of determinations of paleotemperatures in the Upper Cretaceous seas according to the calcium-magnesium ratio index and according to mass-spectrometric data on oxygen isotopes in the rostrums of belemnitellids). Acad. Nauk SSSR, Doklady, v. 175, n. 2, p. 450-451.

Berlin, T. S., and Khabakov, A. V., 1970: Rezul'taty sravneniya Ca/Mg otnosheniy i temperatur po izotopam O^{18}/O^{16} v rostrakh yurskikh i rannemelovykh belemnitov (Results of the comparison of Ca/Mg ratio and temperatures according to O^{18}/O^{16} isotopes in rostra of Jurassic and early belemnites), English summary. Geokhim. Acad. Nauk SSSR, n. 8, p. 971-978.

Berlin, T. S., Kiprikova, Ye. L., Naydin, D. P., Polyakova, I. D., Saks, V. N., Teys, R. V., and Khabakov, A. V., 1970: Nekotoryye problemy paleotemperaturnogo analiza, po rostram belemnitov (Some problems of paleotemperature analysis, on the basis of belemnite rostra), English summary. Akad. Nauk SSSR Sibirsk. Otdelemye Geologiya i Geofizika, n. 4, p. 36-43.

Berlin, T. S., and Khabakov, A. V., 1970: Ca:Mg ratio and O^{18}/O^{16} temperatures for Jurassic and early Cretaceous belemnites, abs. Geochem. Int., v. 7, n. 4, p. 718.

Bowen, R., 1961: Paleotemperature analyses of Mesozoic belemnoidea from Australia and New Guinea. Geol. Soc. Am. Bull., v. 72, n. 5, p. 769-774.

Bowen, R., 1961: Oxygen isotope paleotemperature measurements on Cretaceous Belemnoidea from Europe, India and Japan. J. Paleon., v. 35, n. 5, p. 1077-1084.

Bowen, R., 1962: Paleotemperature analyses of Jurassic Belemnoidea from East Greenland. Experientia, v. 28, n. 10, p. 438-439.

Combemorel, R., Evin, J., and Packiaudi, C., 1973: Mésures de paléotempératures sur les Bélémnites du Crétace inférieur du Sud-Est de la France (Paleotemperature measurements on belemnites from the lower Cretaceous of southeastern France), abs. Réun. Annu. Sci. Terre, Programme Résumés, p. 138.

Eichler, R., and Ristedt, H., 1966: Isotopic evidence of the early life history of *Nautilus pompilus* (Linné). Science, v. 133, n. 3737, p. 734-736.

Epstein, S., Buchsbaum, R., Lowenstam, H. A., and Urey, H. C., 1953: Revised carbonate-water isotopic temperature scale. Geol. Soc. Am. Bull., v. 64, p. 1315-1326.

Fritz, P., 1965: O^{18}/O^{16} isotopenanalysen und paleaotemperaturbestimmungen an Belemniten aus des Schwäb (Paleotemperature determinations based on oxygen isotopic analyses on belemnites from the Jurassic beds of Swabia), English summary. Geol. Rundsch., v. 54, n. 1, p. 261-269.

Germany

Jordan, R., and Stahl, W., 1971: Isotopische palaeotemperaturbestimmungen au jurassischen ammoniten und grundsaetzliche voraussetzungen fuer diese methode (Isotopic paleotemperature determinations in Jurassic ammonites and the fundamental assumptions of this method), English and French summaries. Ger., Bundesanst. Bodenforsch, Geol. Jahrb., v. 89, p. 33-61.

Germany

Kattennegger, W., 1967: Paleotemperaturbestimmungen an aragonitischen dibranchiantenrostren des Trias (Paleotemperature determinations on aragonitic dibranchiate rostrums from the Triassic). Naturwissenschaften, v. 54, n. 19, p. 515.

Kunz, I., 1973: Sauerstoffisotopen-temperaturmessungen an Jura-sedimenten im nordteil der DDR (Temperature measurements, with oxygen isotopes, of Jurassic sediments in northern East Germany), Russian and English summaries. Z. Angew. Geol., v. 19, n. 1, p. 21-27.

Longinelli, A., 1969: Oxygen-18 variations in belemnite guards. Earth Planet. Sci. Lett., v. 7, n. 2, p. 209-212.

Naydin, D. P., Teys, R. V., and Zadorozhnyy, I. K., 1964: Nekotoryye novyye dannyye o temperaturah maastrikhtskikh basseynov Russkoy platformy i sopredel'nykh oblasti po izotopnomu sostave kisloroda v rostrakh belemnitov (Some new data on temperatures of the Maastrichtian basins of the Russian platform and adjacent areas according to the isotopic composition of the oxygen in belemnite rostrums), English abstract. Geokhim., n. 10, p. 971-979.

Naydin, D. P., Teys, R. V., and Zadorozhnyy, I. K., 1966: Izotopniyye paleotemperatury verkhnego mela Russkoy platformy i drugikh rayonov SSSR (Isotopic paleotemperatures of the Upper Cretaceous of the Russian platform and other regions of the USSR), English summary. Geokhim., n. 11, p. 1286-1299.

Spaeth, C., Hoefs, J., and Vetter, U., 1971: Some aspects of isotopic composition of belemnites and related paleotemperatures. Geol. Soc. Am., Bull., v. 82, n. 11, p. 3139-3150.

Stahl, W., and Jordan, R., 1969: General considerations on isotopic paleotemperature determinations and analyses on Jurassic ammonites. Earth Planet. Sci. Lett., v. 6, n. 3, p. 173-178.

East Germany

Stevens, G. R., and Clayton, R. N., 1971: Oxygen isotope studies on Jurassic and Cretaceous belemnites from New Zealand and their biogeographic significance. New Zealand J. Geophys., v. 14, n. 4, p. 829-897.

Tan, F. C., Hudson, J. D., and Keith, M. L., 1970: Jurassic (Callovian) paleotemperatures from Scotland. Earth Planet. Sci. Lett., v. 9, n. 5, p. 421-426.

Teys, R. V., Chupakhin, M. S., and Naydin, D. P., 1957: Opredeleniye paleotemperatur po isotopnomy sostavu kisloroda v kal'tsite rakovin nekotorykh melovykh iskopayemykh Kryma (Determination of paleotemperatures according to the isotopic composition of oxygen in calcite of shells of some Cretaceous fossils from Crimea). Geokhim., n. 4, p. 271-277.

Teys, R. V., Naydin, D. P., Zadorozhnyy, I. K., and Stolyarova, S. S., 1964: O standarte dlya opredeleniya paleotemperatur po izotopnomu sostavu kisloroda organogennogo kal'tsita (On the standard for determining paleotemperatures from the isotopic composition of the oxygen of organogenic calcite), English summary. Geokhim., n. 2, p. 102-109.

Teys, R. V., 1968: Opredeleniye temperatur drevnikh morey (Determination of the temperature of ancient seas). Akad. Nauk SSSR, Vestn., n. 10, p. 28-32.

Siberia

Teys, R. V., Naydin, D. P., and Zadorozhnyy, I. K., 1969: Izotopnyy sostav kisloroda $CaCO_3$ rostov verkhnemelovykh belemnitov i vmeshchayushchikh porod (Oxygen isotope composition of $CaCO_3$ in upper Cretaceous belemnite rostra and enclosing rocks), English summary. Geokhim., n. 1, p. 33-39.

Teys, R. V., and Naydin, D. P., 1969: Bestimmung der Jahresmitteltemperaturen und der jahreszeitlichen temperaturen an hand der sauerstoffisotopen-verteilung im kalzit von marinen fossilen der Russischen tafel (Determination of average annual and seasonal temperatures by means of the oxygen isotope distribution in the calcite of marine fossils of the Russian platform), English and Russian summaries. Berlin, Geologie, v. 18, n. 9, p. 1062-1071.

Tourtelet, H. A., and Rye, R. O., 1969: Distribution of oxygen and carbon isotopes in fossils of late Cretaceous age,western interior region of North America. Geol. Soc. Am. Bull., v. 80, n. 10, p. 1903-1922.

Urey, H. C., Lowenstam, H. A., Epstein, S., and McKinney, C. R., 1951: Measurement of paleotemperatures and temperatures of the Upper Cretaceous of England, Denmark, and the southeastern United States. Geol. Soc. Am. Bull., v. 62, n. 4, p. 399-416.

D. BIOGENIC SILICA

Labeyrie, L., 1972: Composition isotopique de l'oxygène de la silice biogenique (Oxygen isotope composition of biogenic silica). Acad. Sci. Comptes Rendus, Sér. D, v. 274, n. 11, p. 1605-1608.

Labeyrie, L., and Duplessy, J. C., 1973: Géochemie isotopique de la silice biogénique; détérmination d'une nouvelle échelle de paléotempérature (Isotopic geochemistry of biogenic silica; determination of a new scale of paleotemperatures), English summary. Rev. Géogr. Phys. Géol. Dyn., v. 15, n. 5, p. 511-521.

Labeyrie, L., 1974: New approach to surface seawater palaeotemperatures using O^{18}/O^{16} ratios in silica of diatom frustules. Nature, v. 248, n. 5443, p. 40-42.

Mopper, K., and Garlick, G. D., 1971: Oxygen isotope fractionation between silica and ocean water. Geochim. Cosmochim. Acta, v. 35, n. 11, p. 1185-1187.

E. GLACIAL ICE

Dansgaard, W., Johnson, S. J., Moller, J., and Langway, C. C., Jr., 1969: One thousand of climatic record from Camp Century on the Greenland Ice Sheet. Science, v. 166, n. 3903, p. 377-381.

Dansgaard, W., and Tauber, H., 1969: Glacier oxygen-18 content and Pleistocene ocean temperatures. Science, v. 166, n. 3904, p. 499-502.

Dansgaard, W., Johnson, S. J., Clausen, H. B., et al., 1970: Ice cores and paleoclimatology; with discussion (<u>in</u> Radiocarbon variations and absolute chronology, Nobel Symposium, 12th, Uppsala, Proceedings). New York, John Wiley and Sons, p. 337-348; discussion, p. 348-351.

Greenland

Dansgaard, W., Johnson, S. J., Clausen, H. B., et al., 1973: Stable isotope glaciology. Medd. Groenland, v. 197, n. 2, 53 p.

Antarctica and Greenland

Dansgaard, W., Johnson, S. J., Clausen, H. B., and Langway, C. C., 1973: Time scale and ice accumulation during the last 125,000 years as indicated by Greenland O^{18} curve. Geol. Mag., v. 110, n. 1, p. 81-82.

Devereux, I., 1967: Oxygen isotope paleotemperature measurements on New Zealand Tertiary fossils. New Zealand J. Sci., v. 10, n. 4, p. 998-1011.

Devereux, I., Hendy, C. H., and Vella, P., 1970: Pliocene and early Pleistocene sea temperature fluctuations, Mangaopari Stream, New Zealand. Earth Planet. Sci. Lett., v. 8, n. 2, p. 163-168.

Emiliani, C., 1969: Amplitude of the Pleistocene climatic cycles at low latitudes and the oxygen isotopic composition of the ice caps, abs. Geol. Soc. Am., Abstr., Program, pt. 7, p. 56-57.

Emiliani, C., 1970: Pleistocene paleotemperatures. Science, v. 168, n. 3933, p. 822-825.

North America and Europe

Epstein, S., Sharp, R. P., and Gow, A. J., 1970: Antarctic ice sheet; interhemispheric climatic implications of stable isotope analyses of Byrd Station deep cores, abs. EOS, Am. Geophys. Union, Trans., v. 51, n. 4, p. 451.

Epstein, S., Sharp, R. P., and Gow, A. J., 1970: Antarctic ice sheet; stable isotope analyses of Byrd Station cores and interhemispheric climatic implications. Science, v. 168, n. 3939, p. 1570-1572.

Epstein, S., Sharp, R. P., and Gow, A. J., 1971: Climatological implications and stable isotope variations in deep ice cores, Byrd Station, Antarctica. Antarctica J., v. 6, n. 1, p. 18-20.

Ericson, D. B., 1957: Lithological descriptions and micropaleontological analysis of Arctic cores (in Age Measurements and other Isotopic Studies on Arctic Materials). Columbia Univ., Lamont Geological Observatory, Contract AF-19(604)-1063, Final Rpt., 32 p.

Hamilton, W. L., and Langway, C. C., Jr., 1968: A correlation of microparticle concentrations with oxygen isotope ratios in 700-year-old Greenland ice. Earth Planet. Sci. Lett., v. 3, n. 4, p. 363-366.

Johnson, S. J., Dansgaard, W., Clausen, H. B., and Langway, C. C., 1970: Climatic oscillations 1200-2000 A.D. Nature, v. 227, n. 5257, p. 482-483.

Greenland

Johnson, S. J., Dansgaard, W., Clausen, H. B., and Langway, C. C., Jr., 1972: Oxygen isotopic profiles through the Antarctic and Greenland ice sheets. Nature, v. 235, n. 5339, p. 429-434.

Langway, C. C., Jr., 1970: Stratigraphic analysis of a deep ice core from Greenland. Geol. Soc. Am. Spec. Paper, n. 125, 186 p.

Langway, C. C., Jr., and Hansen, B. L., 1973: Drilling through the ice cap; probing climate for a thousand centuries (in Frozen Future, A Prophetic Report from Antarctica). New York, Quadrangle Books, p. 189-204.

Lorius, C., Hagemann, R., Nief, G., and Roth, E., 1968: Teneurs en deutérium le long d'un profil de 106 m dans le névé Antarctique; application à l'étude des variations climatiques (Deuterium content on a 106-m profile in Antarctic neve; application to the study of climatic variations). Earth Planet. Sci. Lett., v. 4, n. 3, p. 237-244.

Moerner, N. H., 1972: Time scale and ice accumulation during the last 125,000 years as indicated by the Greenland O^{18} curve. Geol. Mag., v. 109, n. 1, p. 17-24.

Picciotto, E., Deutsch, S., and Aldaz, L., 1966: The summer 1957-1958 at the South Pole; an example of an unusual meteorological event recorded by the oxygen isotope ratios in the firn. Earth Planet. Sci. Lett., v. 1, n. 4, p. 202-204.

Tauber, H., 1970: Glacier O-18 content and paleoclimate, abs. (in American Quaternary Association, 1st Meeting). Seattle, Washington, Amer. Quat. Assoc., p. 130-131.

Thompson, L. G., 1973: Analysis of the concentration of microparticles in an ice core from Byrd Station, Antarctica. Ohio State Univ., Inst. Polar Stud. Rpt., n. 46, 44 p.